T0215032

Superatoms

Superatoms are a growing topic of interest in nanoscience, bringing the physics of electronic structure together with the chemistry of atomically precise clusters. They offer the prospect of materials design based on the targeted tunability of nanoscale building blocks, creating electronic materials that can be used as everything from catalysts to computing hardware.

This book is designed to be an introduction to the field, covering the history of the concept and related theoretical models from cluster physics. It provides an overview of modern theoretical techniques and presents a survey of recent literature, with particular emphasis on the utilisation of these nanoscale building blocks.

It explores the jellium model, shell structure in nuclear physics, and the relationship of these to the solution of the Schrödinger equation for the atom. The subsequent extension into density functional theory enables multiple examples of recent literature studies to be used to demonstrate the key concepts.

This book is an ideal introduction for students looking to build bridges between cluster and condensed matter physics and the chemistry of superatoms, in particular at a graduate level.

Superatoms

An Introduction

Nicola Gaston

CRC Press
Taylor & Francis Group
Boca Raton London New York

CRC Press is an imprint of the
Taylor & Francis Group, an **informa** business

First edition published 2023
by CRC Press
6000 Broken Sound Parkway NW, Suite 300, Boca Raton, FL 33487-2742

and by CRC Press
4 Park Square, Milton Park, Abingdon, Oxon, OX14 4RN

CRC Press is an imprint of Taylor & Francis Group, LLC

ISBN: 978-0-367-76874-4 (hbk)
ISBN: 978-1-032-41722-6 (pbk)
ISBN: 978-1-003-35943-2 (ebk)

DOI: 10.1201/b23295

Typeset in CMR10
by KnowledgeWorks Global Ltd.

Publisher's note: This book has been prepared from camera-ready copy provided by the authors.

Contents

Preface

This book is based on a course of lectures initially delivered as part of the condensed matter programme in the Department of Physics at the University of Auckland. The goal is to provide an understanding of an emergent field of research in cluster physics: that of the concept of 'superatoms', and the utility of this concept for predicting and manipulating the electronic structure and stability of materials.

The content aims to provide a survey of the last decade or so of research on superatoms, in combination with the existing, textbook concepts of cluster physics, encompassing both theory and experiment. In surveying the literature, priority has been given to presenting current knowledge in a consistent and systematic way rather than to providing an exhaustive list of relevant publications. In particular, it needs to be recognised that work on the topic spans both physics and chemistry, and a consistent language for discussing the concepts is needed: I hope this small volume provides a useful introduction in this sense.

Thanks are due to a number of colleagues who have worked with me on this subject over the last years: in particular, I would like to thank Dmitri Schebarchov, James T. A. Gilmour, Doreen Mollenhauer, Julia Schacht, Lukas Hammerschmidt, Juliet Nelson, and Celina Sikorska.

FIGURE 1: Image credit Shutterstock; adapted from image ID: 297886754

CHAPTER 1

What Makes an Atom?

1.1 SUPERATOMS

So – before we get started: what *is* a 'superatom'?

A superatom is a nanoscale collection of atoms, which exhibits the same kind of electronic structure – highly symmetric electronic orbitals that occur in shells with predictable variations in symmetry and energy – as individual atoms. While there are competing uses of the word in the literature, in the most physically simple case, superatoms are constructed from metal atoms, which bond via valence electrons that delocalise over the whole nanoscale cluster of atoms. In essence, one is then left with a collection of fermions confined within a spherical potential – for which it is straightforward that the general physical description of a many-electron atom should apply.

The concept of the atom itself has a long history. In India, the atomic theory is credited to the philosopher Maharishi Kanada, who described atoms as *invisible*, as they are too small for us to see; *eternal*, such that when a material is broken down into its constituent atoms, they still exist and can be refigured into something else; *spherical*, as if they make up everything in the world, they cannot have a directional dependence of their properties, and noted that they must be able to combine with each other to make larger particles (molecules). There were four types of atom proposed to make up all things: earth, air, water, and fire.

The ancient Greek theory of atoms is often more generally cited, if only because it has gifted the English language (amongst others) with the word atom, from '*atomos*', meaning indivisible. The philosopher Democritus and his teacher Leucippus taught, similarly to Kanada, that atoms were indivisible, and eternal, and that the properties of a material depend on the atoms it is made of; however, a distinction is that they did not consider different types of atoms per se, but relied on different arrangements of atoms to provide the observed differences in material properties.

Over two thousand years later, it is generally John Dalton in the early 19th century who is credited with providing empirical evidence for the atomic

DOI: 10.1201/b23295-1

theory, by establishing those strict stoichiometric combinations of different elemental substances are required to produce molecules in which, for example, there is a 2:1 ratio of different types of atoms, such as in H_2O. The subsequent establishment of an array of elements, which differ in weight and chemical properties in a systematic fashion, was then formalised in the second half of the 19th century with the proposal by Dmitri Mendeleev of the periodic table; the synthesis of elements beyond Plutonium (atomic number 94) over the course of the 20th century has now brought us to the completion of a 118 element set of atoms, with the final member of the current periodic table, Tennessine (atomic number 117) synthesised only in 2010.

Each element differs from its neighbours only by the number of charged particles that it contains: positively charged protons in the nucleus, and negatively charged electrons, that are described in general terms as 'orbiting' that nucleus. Neutrons, uncharged particles otherwise similar in mass to protons, also occupy the nucleus and contribute to its stability: the synthetic elements at the end of the periodic table are generally lacking in neutrons, adding to their instability and consequent tendency to radioactive decay. We have, therefore, long since learned to understand atoms as natural assemblies of subatomic particles, and appreciate them as the fundamental building blocks of all matter. Why then, should it matter, that we can mimic their properties at the nanoscale, by constructing 'superatoms'?

In this first chapter, we will look to answer this question, with reference to the limitations of regular atoms, and the characteristics of atoms that determine their ability to combine with each other to form materials: namely, their electronic structure. To do this, we will move straight into a recap of the basic quantum theory of the atom, starting with the Schrödinger equation. This will be useful background for understanding how the mathematical theory leads to similar electronic shell structure in metal clusters, the topic of Chapter 2.

In Chapter 3, the theory will be developed a bit further, to touch on the ideas behind the Jellium model and Density Functional Theory, which are the conceptual and computational bases for analysis of superatomic clusters. The forms of analysis most useful in characterising superatoms are then covered in Chapter 4, with particular reference to the properties of superatoms that matter most.

Chapter 5 covers the experimental evidence for superatomic shell structure in real metal clusters, with a distinction between the generic idea of 'a metal' and the range of actual metallic elements that can be used for their synthesis – with consequent changes in their adherence to simple models. A further challenge to the predictions of our simple models is presented by the existence of competing ideas about superatoms in the literature – most notably, of super-alkalis and superhalogens, defined primarily by their properties – ionisation potentials or electron affinities that outcompete those of any natural atoms. The existence of such species, in particular as small molecules that may not be easily described by the metal cluster-based superatomic concept, is the topic of Chapter 6, which will be followed by the experimental realisation of ligand-protected metal clusters as superatoms in Chapter 7.

Finally, in Chapters 8 and 9, the practical use of superatoms is explored in detail: first through considering the tunability of individual superatoms, and their interactions with each other to form superatomic molecules, and secondly through the consideration of their ability to self-assemble into nanostructured solid-state materials.

1.2 THE SCHRÖDINGER EQUATION: ON THE WAY TO UNDERSTANDING ELECTRONIC STRUCTURE

We wish to understand the properties of superatoms – both as isolated nanostructures, and in the solid state. In order to do this, we need to understand what we mean by electronic structure, firstly as applied to atoms themselves.

All properties of the material can be traced back – eventually – to the properties of the constituent atoms. However, there are over a hundred elements in the periodic table, of which 80 or so can be considered useful due to abundance or stability, and these can be combined in a very large variety of ways. An individual atom has numerous different electronic states – or orbitals – that its electrons have access to, and that can be employed for different types or modes of bonding. The only way, therefore, to find the actual structure of a given combination of atoms is to find the structure of lowest energy – the *ground state* – by solving the Schrödinger equation for that combination of atoms. This leads us to a generalised many-body problem.

1.2.1 The Many-Body Problem in Quantum Mechanics

We can write down the Hamiltonian for a completely general combination of atoms in a given structure, most simply by switching to atomic units with

$$\hbar = e = m_e = 4\pi\varepsilon_0 = 1$$

$$\hat{H} = -\frac{1}{2}\Sigma_i\nabla^2 - \frac{1}{2}\Sigma_\mu\nabla^2 - \Sigma_{\mu,i}\frac{Z_\mu}{|r_\mu - r_i|} + \Sigma_{\mu<\nu}\frac{Z_\mu Z_\nu}{|r_\mu - r_\nu|} + \Sigma_{i<j}\frac{1}{|r_i - r_j|} \quad (1.1)$$

The electrons are labelled with Latin indices, and the nuclei with Greek.

If we consider each term in this equation separately, we quickly see that some are simpler than others. The first term gives us the kinetic energy of the electrons, and the second is that of the nuclei. However, if we are looking for a solution 'in a given structure', the nuclei should not move – this means we can delete the second term in the Hamiltonian (this amounts to the 'clamped-nucleus approximation').

We could, in principle, solve the Schrödinger equation allowing for the nuclei to move, but as they are so much more massive than the electrons, the time scale for their motion is much greater. So we can ignore the coupling of their motion in most cases.

The third term describes the Coulomb force on the nuclei, due to the presence of the electrons, and the corresponding effect of the nuclei on the electrons. The fourth term is due to the effect of the nuclei on each other. In a given structure, we can also ignore these (they have an energetic consequence, but this can be treated as an external parameter).

The final term is the one that leads to all the complications of electronic structure theory. Using the above arguments about the mass of nuclei and our ability to define the positions of nuclei in a given structure, we can define an 'electronic Hamiltonian' (which amounts to a separation of nuclear and electronic degrees of freedom):

$$\hat{H} = -\frac{1}{2}\Sigma_i \nabla^2 - \Sigma_{\mu,i}\frac{Z_\mu}{|r_\mu - r_i|} + \Sigma_{i<j}\frac{1}{|r_i - r_j|} \tag{1.2}$$

The kinetic energy of the electrons, and the Coulomb force on the electrons in the field of the nuclei, can be described exactly (for a 'given structure', of course). But the final term, which describes the effect of each electron on all the other electrons: this is quite intractable, and the origin of the many-body problem, which we can no longer solve exactly. The most general approach for moving forward is to think of each electron moving in a 'mean field' created by all the other electrons: this is the basis of electronic structure theory for molecules and for solids. It is, however, a severe approximation and a lot of work is needed in order to correct this for more accurate computations later, if needed.

1.2.2 Free Electrons in Three Dimensions

If we start by reminding ourselves of the solution of the Schrödinger equation for a free electron:

$$-\left(\frac{\hbar^2}{2m}\right)\left(\frac{\partial^2\Psi}{\partial x^2} + \frac{\partial^2\Psi}{\partial y^2} + \frac{\partial^2\Psi}{\partial z^2}\right) = E\Psi \tag{1.3}$$

with wavefunction Ψ:

$$\Psi = A\sin \mathbf{k_x}x \, \sin \mathbf{k_y}y \, \sin \mathbf{k_z}z \tag{1.4}$$

and wavevector $\mathbf{k}$:

$$\mathbf{k}^2 = \mathbf{k_x}^2 + \mathbf{k_y}^2 + \mathbf{k_z}^2 \tag{1.5}$$

allowing us to find the second derivative needed to solve the Schrödinger equation:

$$\frac{\hbar^2(k_x^2 + k_y^2 + k_z^2)\Psi}{2m} = E\Psi \tag{1.6}$$

which gives

$$E_k = \frac{\mathbf{k}^2 \hbar^2}{2m}. \tag{1.7}$$

This is purely a kinetic energy, which means we can write it in terms of the usual mass and velocity dependence:

$$\frac{m\mathbf{v}^2}{2} = \frac{\mathbf{k}^2 \hbar^2}{2m}, \tag{1.8}$$

from which we can derive the simple relationship

$$m\mathbf{v} = \hbar\mathbf{k}, \tag{1.9}$$

which we can interpret as meaning that $\mathbf{k}$ is a quantum number – introduced to solve the Schrödinger equation, and which labels each possible energy, E_k. In three dimensions there are three quantum numbers, and increasing any of these – corresponding to motion in the x, y, and z directions – will increase the kinetic energy of the electron.

1.3 ATOMS

In the atomic case, finding the relevant quantum numbers – again 3, for a three-dimensional system – takes a little more work. First, we treat the nucleus as fixed, allowing for the energy of the system in the case of the hydrogen atom, for example, to be reduced to the kinetic and Coulomb energies of the electron:

$$\hat{H} = -\frac{1}{2}\nabla^2 - \frac{Z}{r} \tag{1.10}$$

As the electron is moving in a centrosymmetric potential – orbiting the nucleus, to put more crudely in language borrowed from the Bohr model of the atom – the Schrödinger equation is most naturally solved in spherical polar coordinates. Using a separable Ansatz for the time-independent wavefunction:

$$\Psi(\mathbf{r}) = \Psi(r, \theta, \phi) = R(r)\Theta(\theta)\Phi(\phi), \tag{1.11}$$

and the spherical coordinate frame for the Laplacian:

$$\nabla^2 = \frac{\partial^2}{\partial r^2} + \left(\frac{2}{r}\right)\frac{\partial}{\partial r} + \frac{1}{r^2}\left[\frac{\partial^2}{\partial\theta^2} + \cot\theta\frac{\partial}{\partial\theta} + \csc^2\theta\frac{\partial^2}{\partial\phi^2}\right], \tag{1.12}$$

the Schrödinger equation can be put in the form

$$-\frac{\hbar^2}{2mR}\left[\frac{d^2R}{dr^2} + \left(\frac{2}{r}\right)\frac{dR}{dr}\right] - \frac{\hbar^2}{2m\Theta}\frac{1}{r^2}\left[\frac{d^2\Theta}{d\theta^2} + \cot\theta\frac{d\Theta}{d\theta}\right]$$
$$-\frac{\hbar^2}{2m\Phi}\frac{1}{r^2\sin^2\theta}\frac{d^2\Phi}{d\phi^2} - \frac{Z}{r} = E, \tag{1.13}$$

with ordinary derivatives now replacing the partials due to the separability of the wavefunction. This also allows us to rearrange to isolate the terms that are dependent on each variable, for example:

$$
\frac{1}{\Phi} \frac{d^2\Phi}{d\phi^2} =
$$
$$
-\sin^2\theta \left\{ \frac{r^2}{R} \left[\frac{d^2R}{dr^2} + \left(\frac{2}{r}\right) \frac{dR}{dr} \right] + \frac{1}{\Theta} \left[\frac{d^2\Theta}{d\theta^2} + \cot\theta \frac{d\Theta}{d\theta} \right] + \frac{2mr^2}{\hbar^2} \left[E - \frac{Z}{r} \right] \right\},
$$
$$
(1.14)
$$

which we can identify as an eigenvalue equation, as each side of the equality depends on mutually independent variables, requiring that each is equal to a separation constant.

For example, we can identify the magnetic quantum number from the equation for $\Phi(\phi)$:

$$
\frac{d^2\Phi}{d\phi^2} = -m_l^2 \Phi(\phi). \tag{1.15}
$$

Similar reorganisation of the Schrödinger equation for the other angular variable gives:

$$
\frac{d^2\Theta}{d\theta^2} + \cot\theta \frac{d\Theta}{d\theta} - -m_l^2 \csc^2\theta \Theta(\theta) = -l(l+1)\Theta(\theta), \tag{1.16}
$$

which defines the orbital quantum number, and its relationship to the magnetic quantum number, via the usual expressions:

$$
|\mathbf{L}| = \sqrt{l(l+1)}\hbar; \qquad l = 0, 1, 2, \ldots, \tag{1.17}
$$

and

$$
L_z = m_l \hbar; \qquad m_l = 0, \pm 1, \pm 2, \ldots, \pm l. \tag{1.18}
$$

Finally, the radial part of the Schrödinger equation can be written as the radial wave equation:

$$
-\frac{\hbar^2}{2m} \left[\frac{d^2R}{dr^2} + \left(\frac{2}{r}\right) \frac{dR}{dr} \right] + \frac{-l(l+1)\hbar^2}{2mr^2} R(r) - \frac{Z}{r} R(r) = ER(r) \tag{1.19}
$$

which for the hydrogen atom completely determines the electronic energies due to the symmetry of the potential. Finding a solution to the radial wave equation not only finds us our third quantum number, and the energies it determines, but also the wavefunctions.

1.3.1   The Hydrogen Atom

The angular parts of the wavefunctions that satisfy the Schrödinger equation for the hydrogen atom are given by the spherical harmonic functions, usually written as $Y_l^{m_l}(\theta, \phi)$. The total wavefunction can then be written as

$$
\Psi(\mathbf{r}) = R(r) Y_l^{m_l}(\theta, \phi). \tag{1.20}
$$

The radial wave equation is simplified, first by noting that

$$\frac{d^2(rR)}{dr^2} = r\left[\frac{d^2R}{dr^2} + \left(\frac{2}{r}\right)\frac{dR}{dr}\right] \tag{1.21}$$

which allows us to identify an effective one-dimensional function, $g(r) = rR(r)$, for which we can write a quasi-one-dimensional version of the Schrödinger equation for this three-dimensional problem, effectively by integrating over all possible values of θ and ϕ. Secondly, we can define an effective potential,

$$U_{\text{eff}} = \frac{-l(l+1)\hbar^2}{2mr^2}R(r) - \frac{Z}{r} \tag{1.22}$$

resulting in a simplification of the Schrödinger equation to:

$$-\frac{\hbar^2}{2m}\frac{d^2g}{dr^2} + U_{\text{eff}}(r)g(r) = Eg(r) \tag{1.23}$$

which has solutions

$$E_n = \frac{e^2}{2a_0}\left\{\frac{Z^2}{n^2}\right\}. \tag{1.24}$$

In essence, it is Equation 1.24 that introduces us to the concept of *electronic shells*. The Schrödinger equation operates as an eigenvalue equation, with an eigenvalue – the energy – corresponding to each eigenfunction of the Hamiltonian – the possible wavefunctions of the electron. These wavefunctions define a spectrum of possible states that the electron in the hydrogen atom may adopt, and thereby produces a spectrum of possible energies, E_n. These energies only depend on the principal quantum number n, in this case, with the relationship between the principal quantum number and the angular momentum quantum number as defined above, being

$$l = 0, 1, ..., n - 1. \tag{1.25}$$

This relationship between the quantum numbers for the hydrogenic wavefunctions means that we can write the hydrogenic spectrum as a series of electronic shells, using numbers for the principal quantum number n and the usual spectroscopic notation for $l = 0, 1, 2, 3$ of the symbols s, p, d, f, as

$$1s^2, |2s^2, 2p^6, |3s^2, 3p^6, 3d^{10}, |4s^2, 4p^6, 4d^{10}, 4f^{14}, ... \tag{1.26}$$

where the superscript indicates the number of electrons that can occupy each orbital or (sub)shell. In the hydrogen atom, the fact that the energy only depends on n means that we can group together subshells of orbitals with different angular momentum characters, and use vertical lines to indicate the boundaries between energetic shells, within which all wavefunctions correspond to the same energy eigenvalue. In atoms with more than one electron, the energy does depend on the angular momentum of the wavefunction because the interaction energy of two electrons will depend on the angle between them. This energetic degeneracy is then broken, leading to the identification of what we consider subshells here as real electronic shells with distinct energies. We will see how this arises for many-electron atoms in the following.

1.3.2 The Helium Atom

The 'obvious' proposal for the wavefunction of the helium atom was given by Hartree: that the product of the two individual hydrogen atom orbitals would be a sensible first approximation:

$$\Psi(\mathbf{r}_1, \mathbf{r}_2) = \psi_1(\mathbf{r}_1)\psi_2(\mathbf{r}_2) \tag{1.27}$$

In this way, we base our Ansatz for the wavefunction on the known solution for the H atom, and the 'Hartree product' Ansatz that provides that the individual energies must add, and the individual wavefunctions should be multiplied – but we antisymmetrise this wavefunction, by calculating a determinant. For example, for He:

$$\Psi = \frac{1}{\sqrt{2}}\{\psi_1(1)\psi_2(2) - \psi_1(2)\psi_2(1)\}. \tag{1.28}$$

This antisymmetrisation is necessary for any wavefunctions that describes a fermion, such as an electron.

1.3.3 Generalised Many-Electron Atoms

More generally, we can write many fermion wavefunctions as a combination of individual wavefunctions, so long as it is antisymmetrised:

$$\Psi = \frac{1}{\sqrt{N!}}\text{Det}\{\psi_1(1)\psi_2(2)\psi_3(3)\psi_4(4).....\psi_N(N)\} \tag{1.29}$$

The solution of the Schrödinger equation for a many-electron atom is more complicated, and we need not go into advanced methods of solution here; we will however revisit what is possible here when we come back to discuss methods for the description of real clusters later, where the use of Density Functional Theory for this purpose will be introduced. Up to this point, the approximations we have used include the clamped nucleus approximation (the kinetic energy of the nuclei can be ignored) and the Born-Oppenheimer approximation (the nuclear charge enters only parametrically into the electronic Hamiltonian, and thus the electrons can be thought of as moving independently of the nuclei up until the point where we have to consider that the nuclei are able to move on the electronic potential energy surface). The final approximation, which is responsible for removing much of the complexity of the many-body wavefunction, is the pseudopotential approximation.

The pseudopotential approximation is being used implicitly every time that we refer to the positions of the 'ions' in condensed matter physics, rather than the positions of the nuclei: we are thinking of the valence electrons of the atoms as electronically active and therefore relevant to our calculations, while the core electrons are assumed to stick to the nuclei and occupy essentially the same orbitals that they would in an isolated atom. It turns out that this is a pretty good approximation, and we can describe most of the core electrons by an effective potential that shields the valence electrons from the nuclear charge. This approximation is of particular relevance to the concept of superatoms, where it is only the valence electrons of the substituent atoms that contribute to the superatomic electronic states.

1.4 THE LIMITS OF THE PERIODIC TABLE

There are two broad conclusions we can draw on the basis of the background above.

1. The 'orbital' picture that is used extensively in chemistry to interpret and predict the properties of different atoms and their ability to bond with each other, is based directly on the Hartree Ansatz. This means that all atoms are considered to have electrons 'in orbitals' (to use the language of chemistry) or 'described by single-particle wavefunctions' (to more precise in quantum mechanical terms), that are only perturbations of the original orbitals/wavefunctions that we obtained as solutions to the Schrödinger equation for the hydrogen atom.

2. The quantum mechanical description of the properties of atoms, according to Equation 1.23, depends only on an effective potential, which is spherically symmetric. While we recognise that repulsion between electrons has the potential to break this spherical symmetry, the hydrogen atom orbitals that are good first approximations for many-electron atoms must remain a good first approximation for the behaviour of any fermion moving in a spherically symmetric potential.

As shown in Fig. 1.1, the similarity between the hydrogenic orbitals and the wavefunctions of the outermost electron of each elemental atom – the outermost valence electron – means that the periodic table can be classified in groups, characterised by the nature of that wavefunction. The periodicity of the elements arises from the repetition of the angular momentum quantum number l, as the principal quantum number n is increased. Each atom that has the same valence electron character has the same symmetry of its valence electron wavefunction, leading to an ability to create chemical bonds of the same kind. Thus the s-block elements are considered 'good' metals, while the p-block elements become instead non-metallic with increasing number of p-electrons. The d-block metals have the ability to have interesting magnetic properties, as the higher angular momentum wavefunctions are contained closer to the nucleus, meaning that d-electrons are less likely to participate in bonding, and are more likely to remain unpaired, conferring the potential for magnetic properties to emerge. These are the broad trends, but we will investigate in the following chapters the ways in which variation arises within each of these categories.

At the start of this introductory chapter, we considered the origins of the atomic theory and the scientific progress that has led to our current understanding of a periodic system of elements, of which 118 have been demonstrated to exist, and placed in the periodic table, according to their quantum

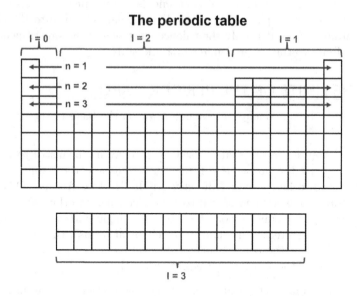

FIGURE 1.1: The relationship between the structure of the periodic table and the quantum numbers that appear in the solution to the Schrödinger equation for the hydrogen atom. The blocks labelled $l = 0, 1, 2, 3$ for the angular momentum quantum number correspond to the atoms that have outermost electrons in an s, p, d, or f electron shell.

numbers. This periodic system is greatly extended beyond the ancient Greek concept of a single type of atom, or even the Indian philosophy within which there were four: however, it has limits, which can be categorised in three ways.

Firstly, the nuclear stability of each element limits its presence on Earth, and its utility for the design of practical materials. The stability of each nucleus depends on its number of protons and neutrons but generally declines with size – this is primarily due to the very small size of the nucleus, which brings the positively charged protons into extreme proximity, leading to massive electrostatic repulsion, indicated by the naming of the strong nuclear force which overcomes this repulsion to hold the nucleus together. As nuclei become larger, more neutrons, which contribute to the strong interaction together with the protons, are required to hold the nucleus together. It is a systematic weakness of the methods of synthesis of the heavy elements, which must be created by the fusion of lighter nuclei, that the resulting isotopes are neutron deficient. However, even improvements resulting in the synthesis of heavier isotopes are unlikely to ever lead to the formation of indefinitely stable nuclei, and thus, we can say that there is an effective, practical limit to the periodic table, resulting from the radioactivity of the heavy elements.

A second question is whether there is a limit due to electronic stability. The Dirac equation, that supersedes the Schrödinger equation for heavy elements where the influence of relativity becomes important, predicts that the innermost electron has an energy

$$E_0 = mc^2 \left[1 - \left(\frac{Z}{137} \right)^2 \right]^{\frac{1}{2}}. \tag{1.30}$$

This energy becomes unphysical at atomic number $Z > 137$, suggesting a limit to the electronic stability of the atom at atomic number 137. However, this limit is extended up to around $Z = 170$ in models that take into account the finite size of the nucleus [1].

The third limitation on the periodic table arises less from the question of physical stability (nuclear or electronic) and is more a limit on our ability to predict or anticipate the properties of novel elements. The electronic configurations of the elements up to atomic number 172 have been calculated [2], and the placement of each element in an extended periodic table has been predicted, according to the symmetry, or angular momentum character, of the outermost valence electron. Some difficulty arises because as atoms grow in size, the energetic spacing between electron shells is reduced, and the stability associated with the completion of an electron shell is also reduced. This means that the clarity of the periodic trends seen for lighter elements is unlikely to be reproduced for heavier elements, even if they can be synthesised. In addition, there are changes to the expected ordering of the different shells that arise due to the influence of relativity, which impacts on the energetics of heavier atoms due to the increased velocity of the innermost electron that experiences the Coulomb force of a nucleus with a charge up to two orders of magnitude greater than its own. Due to the interrelationship of the single-electron orbitals, required due to the orthogonality of the wavefunctions that are required for them to be eigenfunctions of the Hamiltonian, this has energetic consequences for even the valence electrons, and as these consequences are different for shells of different angular momentum characters, it can change the nature of the outermost electron shell: for example, s-shells are contracted due to relativity, while d-shells expand due to the improved shielding of the nucleus by the s-electrons.

In summary: if we are to look for potential building blocks for the design of novel materials, we cannot look to an every increasing array of elements any more, but only to their combinations. However, the ability to design superatoms, based on existing atoms combined in ways that reproduce the symmetry needed for the atom-like electron shells to be reproduced, offers a new concept for materials design. Their construction from combinations of elements also offers the opportunity for their electronic structure to be tuned, through minor, targeted modifications of their composition that take advantage of their predictable electronic shell structure.

Bibliography

[1] Frederick G. Werner and John A. Wheeler. Superheavy nuclei. *Physical Review*, 109(1):126–144, 1958.

[2] Pekka Pyykkö. A suggested periodic table up to Z ≤ 172, based on Dirac-Fock calculations on atoms and ions. *Physical Chemistry Chemical Physics*, 13(1):161–168, 2011.

CHAPTER 2

Electronic Shell Structure: The Case of Ideal, Simple Metal, Clusters of Atoms

Electronic shell structure in metal clusters was first studied in the early 1980s with the advent of experimental cluster sources able to size-select for numbers of atoms, in particular for sizes below 100 atoms. This work was motivated by the increased miniaturisation of electronic devices already by then apparent [1].

It is worth considering that metal clusters themselves have been used over a much longer timescale, however: in the Middle Ages, in Europe, stained glasses were produced in particular colours due to the incorporation of small metal clusters (notably gold) of controllable size, to exploit their size-dependent optical properties. In fact it was Rayleigh who recognised that the scattering of light by small metal particles explained the colours of stained glasses. However, an accurate description of the electronic structure of metal clusters – too large to be molecules, and yet too small to be treated as macroscopic metal, via band theory – required significant experimental and theoretical effort.

2.1   METAL CLUSTERS

A cluster   is a particle containing a countable number of atoms. The word includes, in principle, everything from the diatomic molecule up to aggregates of hundreds of thousands of atoms. In the current text, the word cluster is reserved for the smaller particles in this size range, while larger aggregates for which the electronic structure can be extrapolated from the bulk material, are sometimes distinguished by being referred to as nanoparticles. The boundary between a cluster and a nanoparticles is, however, not a well-defined one.

DOI: 10.1201/b23295-2

2.2   EXPERIMENTAL EVIDENCE OF MAGIC NUMBERS

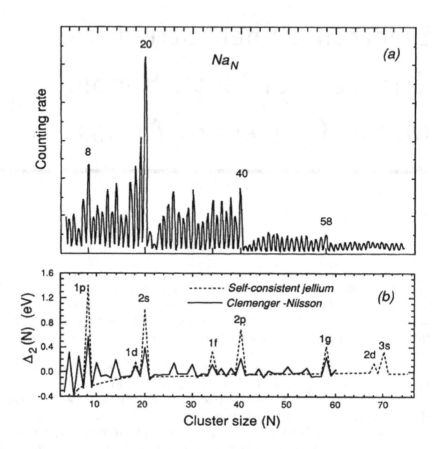

FIGURE 2.1: The sodium cluster abundance spectrum first produced by Knight et al. [2]. Image reproduced with permission.

In 1984, the work of Knight et al. [2] demonstrated the thermodynamic importance of electronic shell structure in a cluster abundance spectrum (see Fig. 2.1). These clusters are formed from a gas of sodium atoms produced by heating a sodium source; as the gas is thermalised, the relative abundances of those clusters that have aggregated from different numbers of atoms, and that are therefore of different sizes, must reflect their relative stabilities.

The numbers for which peaks are seen provide evidence of electronic shell structure: particular numbers of electrons which provide stability in the same way as the closing of electron shells in a noble gas atom. In regular atoms, these closed shells occur at the right hand side of the periodic table, just before the start of a new electron shell, and thus the atoms He, Ne, and Ar correspond to the electron configurations $1s^2$, $[\text{He}]2s^22p^6$, and $[\text{Ne}]3s^23p^6$, with a total of 2, 10, and 18 electrons. The equivalent numbers are referred to as 'magic numbers' in the cluster physics literature, however, the numbers themselves are different, following an electron shell pattern that goes as

$$1S^2, 1P^6, 1D^{10}, 2S^2, 1F^{14}, 2P^6, 1G^{18}, 2D^{10}, 1H^{22}, 3S^2, 2F^{14}, 3P^6, 1I^{26}, ...$$

$$(2.1)$$

From this point onwards, capital letters will be used to refer to super-atomic electron shells (or their corresponding spherical harmonic functions) while small letters will be used for atomic orbitals (and their functions). It becomes immediately evident from the experimental data that the relationship between quantum numbers n and l is not the same in the case of metal clusters, and therefore that higher angular momentum functions are more readily accessible at lower electron counts. One should also note that the electronic shells that occur at 20 electrons, with the closing of the $2S$ shell, or at 40 electrons, with the closing of the $2P$ shell, are sometimes referred to as 'electronic magic numbers' as they result from electron counting, while 'geometric magic numbers' have also been observed due to occur due to geometric factors (the perfect icosahedral symmetry obtainable by a cluster of 13 atoms, in which a single central atom is surrounded by 12 equivalent surface atoms, is an example). Sodium is a very good, free-electron like metal, due to its single valence s-electron, and therefore the number of valence electrons to consider is the same as the number of atoms in the case of these early experiments.

In the bottom panel of Fig. 2.1, the predictions of two electron shell models are presented, and it is straightforward to confirm that the features predicted match very well with experiment. (The Clemenger-Nilsson model takes into account the additional fact that these clusters are not all perfectly spherical.) This corresponds to a simple model of valence electrons confined within an effective potential created by the background ionic cores (see Fig. 2.2 for a schematic).

However, the shape of the confining potential is subject to the internal structure of the metal cluster, which is one aspect that complicates these electron shell-models in comparison to the electronic structure of the atom. In fact, the situation is much more directly related to the shell structure of nuclei, and this is the observation that finally enables us to make progress towards constructing a proper theoretical model for these clusters. Fig. 2.2 contains a sketch of the relative energies of the different quantum states, according to their principal and angular quantum numbers n and l, as the confining potential is changed from the simple harmonic potential to a squarer well potential.

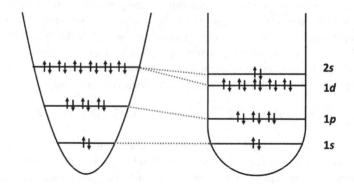

FIGURE 2.2: A schematic of the different quantum states the valence electrons may occupy within a confining potential, according to their principal and angular quantum numbers n and l. While an ideal harmonic potential preserves the energetic degeneracy of wavefunctions with the same number of nodes (radial + angular), any non-ideal potential will break this energetic degeneracy.

2.3 THE EFFECTIVE POTENTIAL

2.3.1 Isotropic Potentials

In contrast to atoms, in which the Coulomb force is unambiguously defined, the nuclear potential is less clearly described and must be approximated. On the other hand, the use of effective potentials in describing the nuclear confining potential may be readily transferred to the description of valence electrons in metal clusters, following the early models of Maria Goeppert Mayer [3].

We consider a cluster of atoms in which the atoms are regularly distributed in space. The density at the centre of the cluster is not expected, necessarily, to be different than anywhere else in the cluster: this assumption of uniform density allows us to say the potential, $U(r)$, is not singular at $r = 0$. Using a second assumption, of spherical symmetry, we can impose the condition that

$$\left(\frac{dU}{dr}\right)_{r=0} = 0. \tag{2.2}$$

The potential must go to zero at the surface of the cluster, at radius R, meaning that

$$\left(\frac{dU}{dr}\right) > \left|\frac{U}{r}\right| \text{ for } r \to R. \tag{2.3}$$

The square well potential satisfies this constraint in the most trivial way, by requiring that

$$U(r) = \begin{cases} -U_0 = \text{cst} & \text{for } r < R, \\ 0 & \text{for } r > R. \end{cases} \tag{2.4}$$

This corresponds to a case in which the first condition is fulfilled everywhere in the cluster, while the second is fulfilled to the extent that there are infinitely strong forces at the cluster surface. A more realistic potential – or at least, a more gentle choice – is the assumption that the radial change of the potential is linear, leading to the harmonic potential well:

$$U(r) = -U_0[1 - (r/R)^2]. \tag{2.5}$$

The energy levels in the three-dimensional harmonic well are given by

$$E - \hbar\omega[2(n-1) + l] = n_0\hbar\omega \tag{2.6}$$

which is very similar to the one-dimensional harmonic well, but for the fact that now we have degeneracies due to the two quantum numbers, n, and l. (As is the case for atoms, there are no energetic consequences of the third quantum number, m_l, in the absence of magnetic fields.) The integer n_0 can be considered to be the total number of quanta of oscillation, for convenience.

There are two consequences to this. Firstly, if the harmonic potential is deformed in any way, this degeneracy will be lifted. In particular, the levels with higher angular momentum have lower energy for the square well. This can be explained by the greater radial extent of states with higher angular momentum.

The second consequence is that the relationship between n and l is not the same as in regular atomic structure, as was observed above in experiment. For a given value of n_0, we can recognise that

$$l = \begin{cases} \text{even} & \text{for } n_0 = \text{ even, with } l = 0, \ldots, n_0 \\ \text{odd} & \text{for } n_0 = \text{ odd, with } l = 1, \ldots, n_0. \end{cases} \tag{2.7}$$

The ground state of the oscillator, $n = 1$, $l = 0$, is non-degenerate and the 1S level. The first excited state, $n = 1, l = 1$ is a 1P level. In contrast to the hydrogen atom solution to the Schrödinger equation, we do not need to increase the principal quantum number to obtain higher angular momentum states: each oscillator shell (defined by the value of n_0) contains a higher value of l, and at the same time all degenerate states of higher n and lower l.

This model is sufficient to explain the high nuclear stability of He^4, O^{16}, and Ca^{40} isotopes; these correspond to the first three states in the harmonic oscillator potential [3]. The same numbers appear as those in the sodium cluster abundance spectrum above, demonstrating as postulated that the confinement potentials for nucleons in the nucleus and electrons in a metal cluster can be similarly approximated.

2.3.2 Anharmonic and Aspherical Potentials

A key assumption in everything we have discussed so far is that the cluster itself is spherical. This is a reasonable assumption in many cases, generally explained by the tendency of a metallic object to minimise its surface area (true in most cases, but there are some exceptions). However, a vacancy in an electronic shell, or the addition of one extra electron to populate a new shell, can effectively lead to a perturbation away from spherical symmetry, just as it does in atoms.

In addition, the isotropic variation of the potential with radius is an approximation which can be improved through the consideration of anharmonicity (which moves the potential towards, for example, the square well limit). An anharmonic distortion parameter can be added to the harmonic oscillator solutions, to provide predictive corrections. The idea that it is necessary to correct the potential will be revisited in the discussion in Chapter 5, where we discuss ligand-protected clusters, that is, metal clusters that are protected in solution by molecules that are bound to their surface.

Finally, it is worth considering one recent study within which an aspherical potential was explicitly introduced for the description of a 67-atom gold cluster [4]. The box-like structure of this cluster (in this case, protected by phosphine molecules) motivated the use of a model potential to describe the electron shells as existing with a box, 1.2 nm x 0.8 nm x 0.8 nm in size. The excellent agreement obtained illustrates the generality of the concept of electronic shells beyond the essentially spherical shells that we usually consider as superatomic, but required more numerical computation than what we expect to cover here. However, in the next chapter, we will cover the essential ingredients of more precise numerical calculations that can be used to describe – and to predict! – the details of real superatomic systems.

Bibliography

[1] Walt A. De Heer. The physics of simple metal clusters: experimental aspects and simple models. *Reviews of Modern Physics*, 65(3):611–676, 1993.

[2] W. D. Knight, Keith Clemenger, Walt A. De Heer, Winston A. Saunders, M. Y. Chou, and Marvin L. Cohen. Electronic shell structure and abundances of sodium clusters. *Physical Review Letters*, 52(24):2141–2143, 1984.

[3] Maria Goeppert Mayer and J. Hans D. Jensen. *Elementary Theory of Nuclear Shell Structure*. John Wiley and Sons, New York, 1955.

[4] Rosalba Juarez-Mosqueda, Sami Kaappa, Sami Malola, and Hannu Häkkinen. Analysis of the electronic structure of non-spherical ligand-protected metal nanoclusters: The case of a Box-Like Ag67. *Journal of Physical Chemistry C*, 121(20):10698–10705, 2017.

CHAPTER 3

From the Jellium Model to Density Functional Theory

As we have seen in the previous chapter, many of the basic features of the electronic structure of small metal clusters can be described already within the framework of a three-dimensional harmonic oscillator. This is sufficient, already, for us to recognise the essential physics of the superatomic model; however, to assess its utility in the description of superatoms – in particular, if we consider the potential for superatoms to be used as the building blocks of real materials – then we need to be able to describe the details of individual superatoms. In particular, the limitations of the simple harmonic oscillator potential include the following assumptions, all of which can be questioned in real clusters:

- The potential is spherical.

- The potential varies uniformly as a function of radius.

- The surface of the cluster does not introduce distortions in the potential.

- The exact positions of the ionic cores can be neglected.

Taking these assumptions one at a time, we can make progress by considering the simplest possible model which can be tuned to account for the electron density of different systems.

3.1 THE JELLIUM MODEL

Jellium   is a quantum mechanical model which describes electrons interacting in a potential created by a uniform positive background charge. It is also referred to as the uniform electron gas (UEG) or homogeneous electron gas (HEG). The electron density is therefore uniform. The advantage of this model is that it retains much of the simplicity of the harmonic oscillator description,

DOI: 10.1201/b23295-3

while being numerically adjustable to account for different electron densities. It is often used in condensed matter physics to obtain the essential quantum behaviour of the electrons, while ignoring the details of the atomic lattice and structure.

3.2 NUMERICAL CALCULATIONS

To numerically find the energies of the valence electrons in the potential produced by the jellium model, we can variationally optimise (minimise) the total energy of the system, subject to the application of the effective jellium potential:

$$U_{\text{eff}}(r) = \begin{cases} -U_0 = \text{cst} & \text{for } r < R_0, \\ 0 & \text{for } r > R_0. \end{cases} \tag{3.1}$$

In this case, both parameters U_0 and R_0 are found by minimising the total energy equation, which depends on the electron density (ρ) as:

$$E(\rho) = \int U(\mathbf{r})\rho(\mathbf{r})d^3r + \frac{1}{2}\int \frac{\rho(\mathbf{r})\rho(\mathbf{r}')}{|r - r'|}d^3r d^3r' + T_s(\rho) + E_{\text{xc}}(\rho) \tag{3.2}$$

in which the first term integrates over the interactions of the electrons with the background charge, and the second term captures the interactions between electrons, via the density at a point r and that at a second point, r'. $T_s(\rho)$ is the kinetic energy of the non-interacting electrons, which is required when we switch to a description of the energy based only on the electron density, and the final term, $E_{\text{xc}(\rho)}$, is the exchange-correlation energy, which again corrects for the mean-field approximation of the first two terms. We will revisit this expression later, when we encounter the use of Density Functional Theory in section 3.5, based on this formalism. For now, we can note that

$$T_s(\rho) = \Sigma_i < \psi_i| - \frac{1}{2}\nabla^2|\psi_i > \tag{3.3}$$

and

$$\rho(\mathbf{r}) = \Sigma_i|\psi_i(\mathbf{r})|^2. \tag{3.4}$$

To make progress, we need to solve an effective Schrödinger equation in the independent particle approximation:

$$\left[-\frac{1}{2}\nabla^2 + U_{\text{eff}}[\rho, \mathbf{r}]\right]\psi_i(\mathbf{r}) = \varepsilon_i\psi_i(\mathbf{r}). \tag{3.5}$$

All the system dependent variations are captured in the background potential, $U(\mathbf{r})$ in Equation 3.5. Specifically, this background potential is

$$U(\mathbf{r}) = e^2 \int \frac{\rho_p(\mathbf{r}')}{|r - r'|}d^3r' \tag{3.6}$$

with the density of the positive background charge a uniform value: $\rho_p(\mathbf{r'}) = \rho_{p0}$. This uniform background charge can account for the difference between monovalent metals by its relation to the size of the metal atom, defined in terms of its Wigner Seitz radius, r_s: the radius of a sphere whose volume is equal to the mean volume per atom (or more generally, per free electron) in a solid.

Thus we have a single parameter within the jellium model:

$$\rho_{p0} = \left[\frac{4\pi}{3} r_s^3 \right]^{-1}. \tag{3.7}$$

3.2.1 The Variational Principle

In the most general terms, variational methods are simply ways of finding the functions which maximise or minimise the value of a quantity that depends upon those functions. In this case, it is the wavefunctions ψ_i which can be numerically varied to minimise the total energy E, as determined by the dependence of E on the wavefunctions, whether directly (via the kinetic energy term) or indirectly (via the electron density ρ). In physical terms, the rationale for finding the extreme, or lowest, value of the energy, is that this lowest energy state is the ground state, which is what we expect the system to be in at equilibrium.

3.3 THE PSEUDOPOTENTIAL MODEL

In order to assess the validity of the jellium model, we want to use a model for the potential that can provide solutions within the same variational framework as outlined above. The key lack within the jellium model is any description of the positions of the ionic cores, composed of nuclei and the core electrons that remain localised at each atomic centre.

Pseudopotentials are descriptions of the ionic cores, that provide the correct effective potential to reproduce the correct energies for the valence electrons. There must, therefore, be a pseudopotential placed at each point in space at which a nucleus would exist: this provides the internal structure within the cluster that we have ignored until now.

In this case, we can write

$$U_{\text{ps}}(r) = \begin{cases} 0 & \text{for } r < r_c, \\ -\frac{Z}{r} & \text{for } r > r_c. \end{cases} \tag{3.8}$$

with r_c the radius of the core, and Z its net charge. The parameter r_c here should be chosen to produce the correct Wigner Seitz radius, r_s, for the bulk metal.

The introduction of pseudopotentials to represent the realistic structure of the cluster means that we no longer have spherical symmetry, and the solution of Equation 3.5 is considerably more complicated as a result. However, including the correct sum over pseudopotentials as the first term in Equation 3.5 already provides an improved description, even if the spherical definition of the effective potential (Equation 3.1) is retained.

3.4 ELECTRONIC STRUCTURE CALCULATIONS FOR METAL CLUSTERS

An early demonstration of the influence of the ionic cores was provided by Martins et al., in 1980 [1]. In this work, they considered two families of crystal lattices, as model systems for sodium clusters (described by a pseudopotential radius of $r_c = 1.7a_0$, chosen to reproduce the bulk binding energy of sodium). The first had a body-centred cubic lattice structure, which is defined by the three-dimensional lattice vectors (in units of the cubic lattice parameter a:

$$
\begin{aligned}
a_1 &= [-0.5, 0.5, 0.5]a \\
a_2 &= [0.5, -0.5, 0.5]a \\
a_3 &= [0.5, 0.5, -0.5]a
\end{aligned}
$$

within which two lattice sites contain atoms, at $\mathbf{r_1} = (0, 0, 0)$ and $\mathbf{r_2} = (0.5a, 0.5a, 0.5a)$.

The second crystal lattice used was an icosahedral lattice. Icosahedral packing has very high symmetry, but does not pack infinitely in three-dimensions (it is not a Bravais lattice) – it is therefore a useful structure to consider for finite clusters, rather than for extended systems.

3.4.1 The Influence of Lattice Structure

In Fig. 3.1, the calculated ionisation potential (IP) and electron affinity (EA) values of specific clusters are presented as a function of the particle radius, with shell structure emerging immediately out of the jellium model calculations, with the expected ordering: $1S^2 1P^6 1D^{10} 2S^2 \ldots$.

These are calculated as:

$$
IP = E(\text{cluster}) - E(\text{cluster} - e^-) \tag{3.9}
$$

and

$$
EA = E(\text{cluster}) - E(\text{cluster} + e^-) \tag{3.10}
$$

The shell effects evident in these calculations – as labelled in the figure – correspond to the periodic behaviour of normal atoms, and are reproduced with the pseudopotential model (not shown in Fig. 3.1). However, we can recognise that in a simple electrostatic model, the IP and EA vary with the radius of a conducting sphere, R. In fact the primary size-dependence goes as

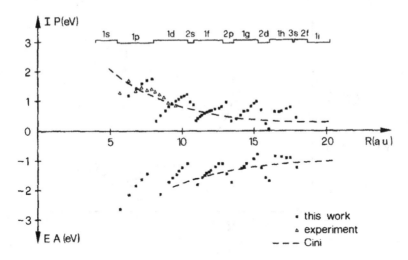

FIGURE 3.1: Calculated ionisation potential (IP) and electron affinity (EA) of sodium clusters, as a function of the particle radius, as calculated with the jellium model in Ref. [1] (labelled 'this work'). Figure reproduced with permission.

$1/R$, based on a simple electrostatic model for a conducting sphere. This is demonstrated in Fig. 3.2, to explain the trend seen in Fig. 3.1.

However, there is an additional reason that non-monotonic variations in the IP or EA may be observed in such finite systems, that is worth comparison with the electronic shell structure seen in Fig. 3.1. The two different lattices considered in this work result in different surface energies $\sigma(R)$, which are calculated from the total energy E by extracting the bulk-like component of the total energy, using the bulk energy per unit volume ε_B:

$$E(R) = \frac{4}{3}\pi R^3 \varepsilon_B + 4\pi R^2 \sigma(R). \tag{3.11}$$

The surface energies are shown in Fig. 3.3, and confirm that the icosahedral clusters have a smaller surface energy in general than the bulk like BCC clusters, which leads to them being more stable in this size range.

However, there are some exceptions: what minimises the surface energy is the ability of the cluster to be spherical, and for a specific number of atoms, the BCC structure can at times adopt a more spherical structure than the icosahedral structure. This observation explains, for example, the bcc cluster at $R = 15.5$ Å, which has 59 atoms and adopts a highly spherical structure at this size. In contrast, the icosahedral structures have high stability at 13 atoms (a central atom surrounded by 12 equidistant surface atoms) and at 55 atoms (the 13 atom cluster surrounded by a shell of equidistant 42 atoms) The

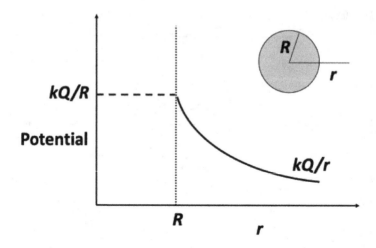

FIGURE 3.2: The potential within a conducting sphere is everywhere equal to kQ/R (k being the Coulomb constant and Q the total charge) as the electric field is zero, and decays outside the sphere with distance r.

55-atom cluster can be seen at around 15 Å. While these icosahedral clusters have a low surface energy by design, as they are constructed in a centrosymmetric fashion, a 59-atom icosahedral cluster would retain a 55 atom core and

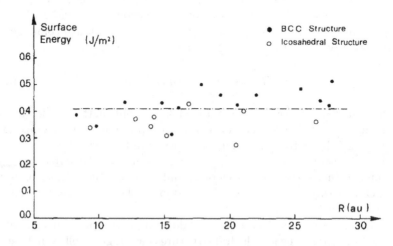

FIGURE 3.3: The surface energies of the different lattice types are compared [1]. Figure reproduced with permission.

have only 4 atoms on the surface of this cluster, and these would be relatively highly destabilised.

These geometric stabilisation factors can lead to non-monotonic behaviours of energies and properties of clusters as the preferred structure can change due to minimisation of surface energy. This can be a factor regardless of the presence of an electronic shell-closing, but it is important to note that electronic shell closings and geometric energetic stabilisation around specific numbers of atoms can also be interdependent. This leads to the idea of a 'doubly magic' cluster: a cluster that has enhanced stability because of both geometric and electronic shell closings. An example is a 13-atom icosahedron with a closed shell number of electrons, which can be arrived at for atoms with more than one valence electron each.

It is useful to note that in general, the radius of the cluster R is approximately related to the number of atoms in the cluster, by

$$R = r_s N^{1/3}. \tag{3.12}$$

3.4.2 The Behaviour of Different Metals

The use of a spherical jellium model can be used to produce effective potentials and corresponding electronic energies for different metals, based on the Wigner Seitz radius – the volume of each atom (or corresponding free electron). Early results obtained by Chou et al., in 1984 [2], are given in Fig.3.4. Here we see the expected consistent behaviour of atoms that belong to a given group in the periodic table.

In contrast to the results presented in Figs. 3.1 and 3.3, here we see the effect of varying only the metal, and instead ignoring the details of lattice structure. In real systems, due to the coexistence of geometric and electronic factors, these effects compete and cooperate, leading to an overall interdependence.

3.5 DENSITY FUNCTIONAL THEORY

Before continuing to a discussion of the detailed differences between different types of superatomic clusters, we need to extend our discussion of the jellium model to the most frequent tool used for atomistic calculation of their properties: that is Density Functional Theory, or DFT. As a *first-principles* theory, the use of DFT assumes no *a priori* knowledge of the system other than the number of electrons, and nuclear charge, attributable to the components of the system. It does however require the use of a functional to extract an energy from a given electron density.

We introduced an equation in Section 3.1 for the total energy, as a function of the electron density: In this case, both parameters U_0 and R_0 are found by minimising the total energy equation, which depends on the electron density (ρ) as:

$$E(\rho) = \int U(\mathbf{r})\rho(\mathbf{r})d^3r + \frac{1}{2}\int \frac{\rho(\mathbf{r})\rho(\mathbf{r}')}{|r - r'|}d^3r d^3r' + T_s(\rho) + E_{\text{xc}}(\rho) \tag{3.13}$$

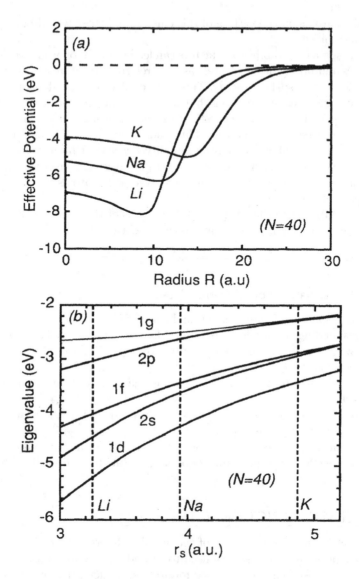

FIGURE 3.4: The jellium model predictions for sodium, potassium, and lithium, are compared, demonstrating the consistent behaviour of alkali metals. Reproduced with permission, after Ref. [2].

with a first and second term that integrate over the interactions of the electrons with the background charge, and with each other, (via the introduction of a second coordinate r'). $T_s(\rho)$ is the kinetic energy of the non-interacting electrons, and $E_{xc}(\rho)$, is the exchange-correlation energy, which again corrects for the mean-field approximation of the first two terms.

Mean-field is the term used to describe this approach to solving a many-body system of interactions, in which each electron is considered to exist against a background of charge determined by both the nuclear positions and by the other electrons. The method used to actually solve equations for the total energy is based on the Hohenberg-Kohn Theorems.

3.5.1 The Hohenberg-Kohn Theorems

Hohenberg-Kohn Theorem One: The external potential is a unique functional of the electron density, $\rho(\mathbf{r})$. As a consequence, the total energy is also a unique functional of the electron density.

To recognise this, we should more properly write our expression for the total energy above as $E[\rho(\mathbf{r})]$, to make explicit that the electron density is a function of position, and thus that the energy is a functional.

The proof of the first theorem is quite simple, and proceeds *reductio ad absurdum*. Let there be two different external potentials, $v_{\text{ext},1}(\mathbf{r})$ and $v_{\text{ext},2}(\mathbf{r})$, that give rise to the same density $\rho_0(\mathbf{r})$. The associated Hamiltonians, $\hat{H}_1$ and $\hat{H}_2$, will therefore have different groundstate wavefunctions, Ψ_1 and Ψ_2, for which both produce the same density $\rho_0(\mathbf{r})$.

The variational principle (introduced briefly in lecture 3) tells us that the ground state energy (the expectation value of the Hamiltonian)

$$E[\rho(\mathbf{r})] = \langle \Psi | \hat{H} | \Psi \rangle, \tag{3.14}$$

can be minimised through variation of the wavefunctions, and if one starts with the 'wrong' wavefunctions for the given Hamiltonian,

$$E_1^0 < \langle \Psi_2 | \hat{H}_1 | \Psi_2 \rangle, \tag{3.15}$$

which leads to

$$\langle \Psi_2 | \hat{H}_1 | \Psi_2 \rangle = \langle \Psi_2 | \hat{H}_2 | \Psi_2 \rangle + \langle \Psi_2 | \hat{H}_1 - \hat{H}_2 | \Psi_2 \rangle \tag{3.16}$$

$$= E_2^0 + \int \rho_0(\mathbf{r})[v_{\text{ext},1}(\mathbf{r}) - v_{\text{ext},2}(\mathbf{r})]\, d\mathbf{r}. \tag{3.17}$$

Expanding the relevant expectation values leads to the contradiction:

$$E_1^0 + E_2^0 < E_2^0 + E_1^0 \tag{3.18}$$

and thereby to the realisation that the groundstate density uniquely determines the external potential $v_{\text{ext}}(\mathbf{r})$. (More accurately, this is true to within an additive constant.)

In physical terms, the electrons (which are free to move) determine the positions of the nuclei in a system, and thereby define all groundstate electronic properties, because $v_{\text{ext}}(\mathbf{r})$ and N completely define $\hat{H}$.

Hohenberg-Kohn Theorem Two: The groundstate energy can be obtained variationally. This is the same as recognising that the density that minimises the total energy is the exact groundstate density.

This proof is also very simple. The electron density, $\rho(\mathbf{r})$ determines $v_{ext}(\mathbf{r})$, while N and $v_{ext}(\mathbf{r})$ determine $\hat{H}$ and therefore Ψ. This ultimately means Ψ is a functional of $n(\mathbf{r})$, and so the expectation value of $\hat{H}$ is also a functional of $\rho(\mathbf{r})$, that is,

$$\hat{H}[\rho(\mathbf{r})] = \langle \psi | \hat{H} | \psi \rangle \tag{3.19}$$

The energy functional $E[\rho(\mathbf{r})]$ alluded to in the first Hohenberg-Kohn theorem above can be written in terms of the external potential $v_{ext}(\mathbf{r})$ in the following way,

$$E[n(\mathbf{r})] = \int \rho(\mathbf{r}) \, v_{ext}(\mathbf{r}) \, d\mathbf{r} + F[\rho(\mathbf{r})] \tag{3.20}$$

where $F[n(\mathbf{r})]$ is an unknown functional of the electron density $\rho(\mathbf{r})$ only. This means that we can write a Hamiltonian for the system as

$$E[\rho(\mathbf{r})] = \langle \Psi | \hat{H} | \Psi \rangle \tag{3.21}$$

to minimise the expectation value that gives the groundstate energy for an electron wavefunction Ψ.

The Hamiltonian can be written as,

$$\hat{H} = \hat{F} + \hat{V}_{ext} \tag{3.22}$$

where $\hat{F}$ is the electronic Hamiltonian consisting of a kinetic energy operator $\hat{T}$ and an interaction operator $\hat{V}_{ee}$,

$$\hat{F} = \hat{T} + \hat{V}_{ee} \tag{3.23}$$

The electron operator $\hat{F}$ is the same for all N-electron systems, so $\hat{H}$ is completely defined by the number of electrons N, and the external potential $v_{ext}(\mathbf{r})$.

The ground-state density of an external potential is known as v-representable. This enables us to define a v-representable energy functional $E_v[\rho(\mathbf{r})]$. Here the external potential $v(\mathbf{r})$ is unrelated to a different density $\rho'(\mathbf{r})$, and

$$E_v[\rho(\mathbf{r})] = \int \rho'(\mathbf{r}) \, v_{ext}(\mathbf{r}) \, d\mathbf{r} + \hat{F}[\rho'(\mathbf{r})] \tag{3.24}$$

and the variational principle allows us to assert that

$$\langle \psi' | \hat{F} | \psi' \rangle + \langle \psi' \hat{V}_{ext} \psi' \rangle > \langle \psi | \hat{F} | \psi \rangle + \langle \psi | \hat{V}_{ext} | \psi \rangle \tag{3.25}$$

where ψ is the wavefunction associated with the correct groundstate $\rho(\mathbf{r})$. This leads to the following integrals:

$$\int \rho'(\mathbf{r})\, v_{\text{ext}}(\mathbf{r})d\mathbf{r} + F[\rho'(\mathbf{r})] > \int \rho(\mathbf{r})\, v_{\text{ext}}(\mathbf{r})d\mathbf{r} + F[\rho(\mathbf{r})] \qquad (3.26)$$

and so the second Hohenberg-Kohn theorem is shown to produce the variational principle as below:

$$E_v[\rho'(\mathbf{r})] > E_v[\rho(\mathbf{r})]. \qquad (3.27)$$

The Hohenberg-Kohn theorems underlie all of modern Density Functional Theory, and are thus very powerful. However, there remain two significant gaps in our understanding:

- We have no idea of the best way to carry out the calculations in practice. The use of a mean-field approach to doing this will be discussed in the next section: using the Kohn-Sham equations, based, pragmatically, on our understanding of how to solve the Schrödinger equation for many-electron atoms using the Independent Particle Approximation.

- While theorem one tells us that there exists a functional that can reproduce the exact energy of the system from the ground state electron density, as that uniquely defines the Hamiltonian and wavefunction, it does not tell us what this functional looks like. The search for the correct functional is one of the remaining serious deficiencies in DFT.

3.5.2 The Kohn-Sham Equations

The Kohn-Sham formulation of DFT is based on a fictitious non-interacting system, such that by construction each electron moves within an effective 'Kohn-Sham' single-particle potential $v_{\text{KS}}(\mathbf{r})$. This enables the implementation of a mean-field approach. The Kohn-Sham method is still exact, in principle, as it has the potential to produce the same groundstate density as the real system. However, to achieve this, we still rely on finding the correct functional.

The second Hohenberg-Kohn theorem enables us to obtain the ground-state energy of a many-electron system variationally – by minimising the energy functional. An important constraint is that the number of electrons N is conserved, which leads to:

$$\delta \left[F[\rho(\mathbf{r})] + \int v_{\text{ext}}(\mathbf{r})\rho(\mathbf{r})d\mathbf{r} - \mu \left(\int \rho(\mathbf{r})\, d\mathbf{r} - N \right) \right] = 0. \qquad (3.28)$$

We can write this as an Euler equation with Lagrange multiplier μ:

$$\mu = \frac{\delta F[\rho(\mathbf{r})]}{\delta \rho(\mathbf{r})} + v_{\text{ext}}(\mathbf{r}) \qquad (3.29)$$

The point of this is that we can then find the function that optimises the ground state energy. The corresponding groundstate wavefunction Ψ_{KS} for the Kohn-Sham system is given exactly by a determinant of single-particle orbitals (wavefunctions) $\psi_i(\mathbf{r}_i)$,

$$\Psi_{KS} = \frac{1}{\sqrt{N!}} \det[\psi_1(\mathbf{r}_1)\psi_2(\mathbf{r}_2)\ldots\psi_N(\mathbf{r}_N)] \tag{3.30}$$

The universal functional $F[\rho(\mathbf{r})]$ is then partitioned into three terms, the kinetic energy $T_s[\rho(\mathbf{r})]$ of a non-interacting electron gas of density $\rho(\mathbf{r})$, and the classical electrostatic (Hartree) energy of the electrons, both of which are known exactly, and an unknown quantity called the exchange-correlation energy $E_{XC}[\rho(\mathbf{r})]$:

$$F[\rho(\mathbf{r})] = T_s[\rho(\mathbf{r})] + E_H[\rho(\mathbf{r})] + E_{XC}[\rho(\mathbf{r})] \tag{3.31}$$

with

$$E_H[\rho(\mathbf{r})] = \frac{1}{2} \int \int \frac{\rho(\mathbf{r})\rho(\mathbf{r}')}{|\mathbf{r}-\mathbf{r}'|} \, d\mathbf{r} \, d\mathbf{r}' \tag{3.32}$$

The exchange-correlation energy E_{XC} captures the difference between the exact and non-interacting kinetic energies and the non-classical contribution to the electron-electron interactions, of which the exchange energy is a part.

The Euler equation above can be rewritten as

$$\mu = \frac{\delta T_s[\rho(\mathbf{r})]}{\delta\rho(\mathbf{r})} + v_{KS}(\mathbf{r}) \tag{3.33}$$

where the Kohn-Sham potential $v_{KS}(\mathbf{r})$ is

$$v_{KS}(\mathbf{r}) = v_{ext}(\mathbf{r}) + v_H(\mathbf{r}) + v_{XC}(\mathbf{r}) \tag{3.34}$$

with the exchange-correlation potential $v_{XC}(\mathbf{r})$,

$$v_{XC}(\mathbf{r}) = \frac{\delta E_{XC}[\rho(\mathbf{r})]}{\delta\rho(\mathbf{r})} \tag{3.35}$$

and the Hartree potential $v_H(\mathbf{r})$,

$$v_H(\mathbf{r}) = \frac{\delta E_H[\rho(\mathbf{r})]}{\delta\rho(\mathbf{r})} = \int \frac{\rho(\mathbf{r}')}{|\mathbf{r}-\mathbf{r}'|} \, d\mathbf{r}' \tag{3.36}$$

The most important point is to see that this expression for the Hartree potential is based on the same density as the ground state density for the non-interacting 'fictitious' Kohn-Sham system. If the observable positions (density) of the electrons is the same, then we must be able to extract the exact energies and properties of the system (in principle).

In practice we obtain the ground state density by solving the N one-electron Schrödinger equations,

$$\left[-\frac{1}{2}\nabla^2 + v_{KS}(\mathbf{r}) \right] \psi_i(\mathbf{r}) = \varepsilon_i \psi_i(\mathbf{r}) \tag{3.37}$$

with Lagrange multipliers (ε_i) now needed for each of the N single-particle states $\psi_i(\mathbf{r})$.

The density is constructed from,

$$\rho(\mathbf{r}) = \sum_{i=1}^{N} |\psi_i(\mathbf{r})|^2 \tag{3.38}$$

The non-interacting kinetic energy $T_S[n(\mathbf{r})]$ is finally given by the usual expression,

$$T_S[\rho(\mathbf{r})] = -\frac{1}{2}\sum_{i=1}^{N} \int \psi_i^*(\mathbf{r})\nabla^2\psi_i(\mathbf{r})\,d\mathbf{r} \tag{3.39}$$

All of this leads to a set of effective one-particle equations, which can be solved iteratively within the Independent Particle Approximation. Since $v_{KS}(\mathbf{r})$ depends on the density, the potential needs to be iteratively updated until the energy converges: this makes this a self-consistent computational procedure.

While N equations have to be solved in Kohn-Sham DFT, the complexity of the calculations only increases in proportion to the number of single-particle equations to be solved. Moreover, although Kohn-Sham theory is exact in principle, it is only ever approximate in practice because of the unknown exchange-correlation functional $E_{XC}[\rho(\mathbf{r})]$. We can define this by working backwards from what we do know, as:

$$E_{XC}[\rho(\mathbf{r})] = T[\rho(\mathbf{r})] - T_S[\rho(\mathbf{r})] + E_{ee}[\rho(\mathbf{r})] - E_H[\rho(\mathbf{r})] \tag{3.40}$$

where $T[\rho(\mathbf{r})]$ and $E_{ee}[\rho(\mathbf{r})]$ are the exact kinetic and electron-electron interaction energies respectively.

Current approximations used for the exchange-correlation functional, and their relative strengths and weaknesses, will be discussed in the context of the systems to be discussed in the following chapters.. While one of the frustrations of modern DFT can be that it feels like there is a zoo of functionals, and no good way to choose between them, it turns out that understanding a few good examples, and their relative performance for ideal metallic, insulating, finite and extended systems, can provide a decent amount of physical intuition about the pros and cons of different choices.

Bibliography

[1] J. L. Martins, R. Car, and J. Buttet. Variational spherical model of small metallic particles. *Surface Science*, 106(1-3):265–271, 1981.

[2] M. Y. Chou, A. Cleland, and M. L. Cohen. Total energies, abundances, and electronic shell structure of lithium, sodium, and potassium clusters. *Solid State Communications*, 52(7):645–648, 1984.

CHAPTER 4

From Density Functional Theory to Properties

The utility of density functional theory lies not only in its ability to calculate the atomistic structures of nanoscale and bulk materials from first principles, but to explain why one particular arrangement of atoms is favoured over another. By analysing the Kohn-Sham states – eigenvalues and eigenfunctions produced by solving the Kohn-Sham equations, in the same way as energies and wavefunctions are found by solving the Schrödinger equation – we can discover how the energies of the electrons change as the atoms are arranged in different ways, and thereby provide some understanding of the forces involved in determining material stability. In the case of superatoms, the extent to which the electronic states map onto the predictions of the jellium model is typically the first indicator of superatomic character.

In the second part of this chapter, the physics that determines the relative accuracy of different functionals will be discussed, with specific reference to the aspects most relevant to the description of the delocalised electronic states found in superatoms.

4.1 ANALYSIS TECHNIQUES FOR ELECTRONIC STRUCTURE OF SUPERATOMS

In the following, we will discuss methods for the interpretation of electronic structure, most of which centre on understanding the density of states (DOS). This is the finite particle equivalent of the band structure that is so critical for understanding the electronic properties of bulk materials and is extensively used in the description of superatoms. We will revisit the use of band structure calculations for the electronic structure of bulk solids in Chapter 9.

DOI: 10.1201/b23295-4

4.1.1 The Density of States

We will briefly recap the concept of the density of states in order to understand how it is used for the analysis of electronic structure in superatoms. The Pauli principle – a 'filling-up principle', in this context – enables us to straightforwardly determine the Fermi energy of a N electron metallic system, where we are considering our conduction electrons to be free electron like, as

$$\epsilon_F = \frac{\hbar^2}{2m}\left(\frac{n_F\pi}{L}\right)^2 = \frac{\hbar^2}{2m}\left(\frac{N\pi}{2L}\right)^2 \tag{4.1}$$

with $N = 2n_F$, n_F being the quantum number of the highest filled state.

In 3D, this becomes a little more complicated, but we can use some simple geometric arguments to make progress. For a given vector $\mathbf{n}$, the three components $\mathbf{n}_x, \mathbf{n}_y, \mathbf{n}_z$ define a particular quadrant of a sphere with volume $V = \frac{\pi}{6}\mathbf{n}^3$. The standing wave solutions to the SE in 1D have

$$\mathbf{k}^2 = \frac{\pi^2\mathbf{n}^2}{L^2} \tag{4.2}$$

so

$$\mathbf{n} = V^{\frac{1}{3}}\frac{\mathbf{k}}{\pi} \tag{4.3}$$

allowing us to write the number of states at a given value of wave vector $\mathbf{k}$ as

$$G(\mathbf{k}) = \frac{\pi\mathbf{n}^3}{6} = \frac{V\mathbf{k}^3}{6\pi^2}. \tag{4.4}$$

In practise we are often interested in how many states we have at a given energy, rather than wave vector. Using the free-electron relationship between E and $\mathbf{k}$,

$$E = \frac{\hbar^2\mathbf{k}^2}{2m} \tag{4.5}$$

we can write that

$$G(E) = \frac{V2mE^{\frac{3}{2}}}{3\pi^2\hbar^3}. \tag{4.6}$$

where we have also doubled the number of states to allow for electrons of opposite spin to have the same (other) quantum numbers (or to'occupy the same orbital')

We can produce a DOS by integrating this expression to get a number per energy interval:

$$\frac{dG}{dE} = \frac{d}{dE}\frac{V2mE^{\frac{3}{2}}}{3\pi^2\hbar^3} \tag{4.7}$$

or

$$g(E) = \frac{V2m^{\frac{3}{2}}E^{\frac{1}{2}}}{2\pi^2\hbar^3}. \tag{4.8}$$

Of particular interest in analysing the DOS produced by a free-electron density:

- the DOS goes as $E^{1/2}$. This is a consequence of the quadratic relationship between E and $\mathbf{k}$ in the free electron model;

- the way the DOS depends on E is called the dispersion;

- the greater the volume, the more states we can include.

In finite systems, or even just in real metals that are imperfectly free-electron like, this scenario needs further consideration. Instead of an extended system, within which we can think of something approximating a continuum of electronic states, we have a finite cluster, with orbitals populated by electrons shared across atoms – but nonetheless, these cluster orbitals are of finite extent, and can be plotted using the energy eigenvalues that are obtained, for example, from a DFT calculation for a given cluster.

In Fig. 4.1, a schematic version of the DOS is given for a superatomic cluster, in which the superatomic $1S, 1P, \ldots$ orbitals are broader than in atomic spectra due to interatomic bonding, but nonetheless remain finite in width. In this picture, the $1S, 1P$, and $1D$ orbitals are shown below the Fermi energy (E_F) and are therefore occupied, while the subsequent $2S$ shell is unoccupied. The integrated area of each peak in the density of states corresponds to the number of electrons in that particular shell (or subshell). Such an 18 electron system as is shown here could correspond to a 18 atom cluster of sodium, for example.

It is less straightforward to identify the angular momentum character of each superatomic orbital than it is to plot the energies, although as mentioned,

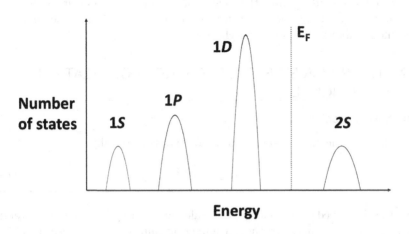

FIGURE 4.1: A schematic of the density of states of a superatomic cluster with 18 electrons.

the peak heights and widths do provide some clue. A very useful technique for assessing the angular momentum character of each superatomic orbital is to project the associated electron density onto a basis of spherical harmonic functions, and assess the numerical overlap.

4.1.2 Spherical Harmonic Projection

The interpretation of the DOS using spherical harmonics projections became particularly useful with the advent of chemically-synthesised, ligand-protected metal clusters for which simply counting the number of metal atoms is insufficient to understand the electronic structure [2]. To do this, spherical harmonic functions are placed at the centre of the cluster, and the overlap coefficients between these functions and the electron density produced with DFT are calculated for each value of l as:

$$c_l(R_0) = \sum_m \int_0^{R_0} \mathbf{r}^2 dr |\psi_{nlm}(\mathbf{r})|^2 \tag{4.9}$$

with

$$\psi_{nlm}(\mathbf{r}) = \int dr Y_l m(\mathbf{r}) \psi_n(\mathbf{r}). \tag{4.10}$$

A sphere of radius R_0 needs to be chosen, within which to integrate the overlap numerically, If chosen to be too large, it can introduce numerical noise and also increase numerical cost significantly; if chosen to be too small, it will artificially truncate the height of the DOS peaks. However, if the projected DOS (PDOS) is then plotted for each value of l, significant value can be added to the interpretation of the electronic structure of these clusters, as shown in Fig. 4.2. Here the angular momentum character of the projected density is portrayed via shading, and the height of the projected density is less than that of the total DOS, as the projection coefficients $c_l \leq 1$. These figures will become familiar in the following chapters.

4.2   THE PHYSICS OF THE EXCHANGE-CORRELATION FUNCTIONAL

The Self Interaction

The electrostatic electron-electron interaction is classically

$$E_H = \frac{1}{2} \int \frac{\rho(\mathbf{r}_1)\rho(\mathbf{r}_2)}{4\pi\epsilon_0 |\mathbf{r}_1 - \mathbf{r}_2|} d\mathbf{r}_1 d\mathbf{r}_2. \tag{4.11}$$

and this is indeed what one gets as calculated with the Hartree wavefunction. The antisymmetry of the wavefunction imposed by the use of a Slater determinant means that the contribution to the potential of the electron itself is removed for the calculation of its own energy – this is a correction for a

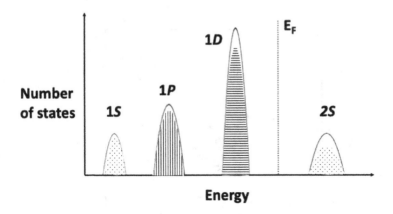

FIGURE 4.2: A schematic of the projected density of states of a superatomic cluster with 18 electrons.

spurious *self-energy*. This self energy is exactly cancelled by the calculation of exchange – which is done exactly within wavefunction theory, in which the use of anti-symmetric wavefunctions allows for the energetic cost of exchanging Fermions to be accounted for. This combination of a product wavefunction Ansatz, due to Hartree, with anti-symmetrisation of the wavefunction to account for particle exchange correctly, is what is known as the Hartree-Fock method for solution of the many-electron wavefunction.

The Exchange Energy

The self-interaction energy is always positive: it increases the energy of localised states and favours delocalisation. The consequences for calculations of band gaps are consistently that the energies of the conduction bands are too low, and thus calculations that include the self-interaction error (generally: all DFT calculations without specific corrections) will predict too low band gaps, and sometimes as a consequence a non-metal can look like a metal (with no band gap) in DFT calculations.

One way to correct for this general problem in DFT is to calculate the Hartree-Fock exchange (which is exact) and add this to correct the band energies.

In condensed phases, the exchange energy also manifests in the Heisenberg Hamiltonian, which is a precursor to the Ising model for spin-exchange and magnetism:

$$H_{\text{Heis}} = -J\Sigma_{i,j} < \mathbf{S}_i \cdot \mathbf{S}_j > \tag{4.12}$$

Here J is the *exchange constant* which is defined as half the difference in energy between the two different eigenfunctions found from the overlap of the

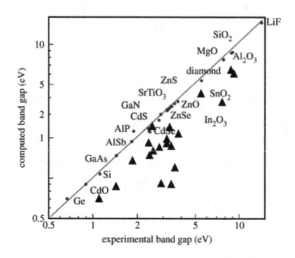

FIGURE 4.3: A comparison of band gaps calculated with DFT with experimental data, reproduced from Ref. [1] with permission. The triangles are for a regular DFT calculation (performed with the PBE functional) while the circular data points have been approximately corrected for the effects of exchange.

two initial wavefunctions: the bonding (spatially symmetric) and antibonding (spatially anti-symmetric) solutions.

The energy of the Ising system is defined then as

$$E_{\text{Ising}} = -\Sigma_{i,j} J_{i.j} < S_i^z S_j^z > \tag{4.13}$$

These models for ferromagnetic interactions between the spins on metal atoms work well for metals where the spins are indeed well localised, but dramatically underestimate the exchange in metals where the electrons are in fact highly delocalised. This competition between localisation and delocalisation is a feature of all electronic structure methods in the sense that the balance between these competing effects is what leads to so much of the particularities of different metals.

This competition between localised states and delocalised states is obviously crucial to treat accurately in considering superatomic electronic shell structure.

The Correlation Energy

While the exchange energy is a consequence of the requirement for antisymmetric wavefunctions for fermions, the correlation energy is the other true many-body energetic contribution. It is sometimes defined (within the single particle, wavefunction picture) as the energy that is missing from the Independent Particle wavefunction (in Hartree-Fock theory): when we solve the SE for independent electrons with a Slater determinant wavefunction according to a self-consistent field procedure, then all we are missing is the instantaneous response of each electron to the exact positions of all the other electrons. The correlation energy is thus a truly non-local effect (which is also why it is hard to calculate).

Within WFT, it is important (and useful!) to note that electron correlation is always a stabilising effect that will lower the total energy of the system. This is important because it leads to the idea that wavefunction based calculations are *systematically improvable*: in contrast, DFT (in which both exchange and correlation energies are approximated) cannot be *systematically* improved, as the exchange-correlation functional can overestimate the necessary correction.

4.3 FAMILIES OF FUNCTIONALS

The Local Density Approximation

The local density approximation is one of the first sensible approximations and is based on the exchange-correlation energy per particle of a uniform electron gas, ϵ_{xc}.

$$E_{xc}^{\text{LDA}}[\rho(\mathbf{r})] = \int \rho(\mathbf{r})\epsilon_{xc}(\rho)d(\mathbf{r}) \tag{4.14}$$

The correlation energy component of this is known analytically only in high and low density limits that correspond to infinitely weak or strong correlation effects. Thus even in the LDA, the functional is not known exactly - modern functionals generally employ the results of quantum Monte Carlo simulations of the UEG at intermediate densities.

The Generalised Gradient Approximation

In the context of real solids, we need to include the variation of electron density between lattice sites as a component of how the electron density affects the total energy. The starting point for this is to require that the functional depends on the gradient of the density, and not just its magnitude.

Despite the change of name, all GGAs are still *local* because the energy still depends only on the local properties of the density (its value or the value of its gradient at $\mathbf{r}$).

Accuracy and Challenges

The biggest issue with modern DFT is that there is no way of knowing, in general, whether your calculated energy is bounded from below, as despite the Hohenberg Kohn theorems, the variational principle only applies *exactly* for the one true functional (which we don't know). Therefore DFT is not systematically improvable in the way that wavefunction based methods are.

The Hohenberg Kohn theorems also only apply to the ground state electron density. Thus the use of DFT is strictly limited to the calculation of ground state properties. The eigenvalues associated with unoccupied bands can be obtained, but these need to be treated with some caution.

The incomplete correction of the self-interaction leads to (in some cases severe) underestimation of the band gap in both LDA and GGA based DFT calculations. In many systems, the self-interaction error is minor for occupied energy levels, structures, and other properties, but for the band gap but it can be a significant problem. The self-interaction energy is always positive and therefore it raises the energy of localised states and leads to increased delocalisation (as discussed previously). This problem can be corrected by the inclusion of 'exact' exchange (based on a HF calculation) but this can still only be an approximate correction (though it does systematically improve band gaps). This leads to the concept of a 'hybrid' functional: a GGA with a percentage of HF exchange included.

Exchange corrections in superatomic shell structure

A nice example is given by nickel clusters, in which the presence of magnetic states created by the d-electrons makes the competition between localised and delocalised states particularly notable. This is presented in Fig. 5.9, for the positively charged ten atom cluster. The gap between the occupied and unoccupied states in the spin up part of the density of states is increased by more than a factor of 2 (over 1 eV) when the exchange interaction is properly included (in the lower panel). More than this, however, it is possible to see how much more localised the d-electron states become when the exchange interaction is included to correct for the self-interaction: the simple characterisations of the electronic structure possible in the top panel, with well defined peaks, become much more complicated in the lower panel. In the lower panel, with the inclusion of exact exchange, the only states that remain so clearly defined are those that are in fact superatomic, and follow the shell closing rules known from the jellium model.

4.4 BEYOND THE INDEPENDENT PARTICLE APPROXIMATION

More accurate computational schemes require consideration of the quasiparticle nature of a screened electron in a many-body system. As we have seen,

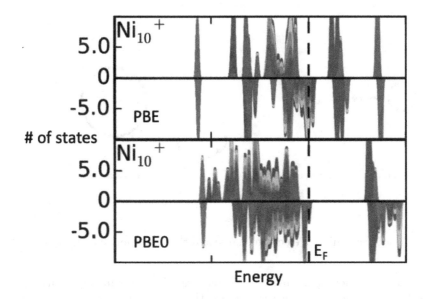

FIGURE 4.4: Comparison between the projected density of states of the Ni_{10}^+ cluster obtained with the PBE (GGA) functional, both without (top) and with (bottom) the inclusion of exact exchange. The energy difference from the Fermi energy (dashed line) to the next energetic state is the band gap, showing the dramatic effect of exact exchange in these cases. The shading indicates the projection coefficients obtained from projecting the Kohn-Sham orbitals onto spherical harmonic functions.

a consequence of the combined failures and limitations of wavefunction based theories and DFT is that we are left, in both cases, with a 'band gap problem' that is quite intractable (though if can be compensated for in an approximate way in each framework).

One way of thinking about the limitations of WFT or DFT, is that the single particle framework limits our ability to think in terms of the quasiparticles that are at times a more natural framework for the description of many-body systems.

In the formalism of second quantisation, we can introduce creation and annihilation operators, $\hat{a}^\dagger$ and $\hat{a}$, which act on the wavefunction to create electrons in specific energy states, or to remove electrons from specific energy states. The relevance of this formalism to condensed matter is most directly made clear by the example of photoexcitations in semiconductors, where an electron that is promoted to an excited state may be described as having been annihilated from one energy level and created in a higher level. (This is also the picture to keep in mind for how correlated wavefunctions are described

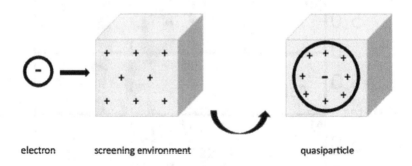

electron   screening environment   quasiparticle

FIGURE 4.5: Screening of electron charge and the formation of a quasiparticle.

to increase accuracy of the many-body picture – by removing electrons from one orbital and creating an electron in another orbital, to describe electronic transitions.)

Thus, if we write the ground state wavefunction for an N-electron system as $|N, 0 >$, we can consider excitations that look like

$$a^\dagger(\mathbf{r}, t)|N, 0 >$$

which produces an $(N + 1)$-electron wavefunction (not necessarily in the ground state).

Because we can describe the coupled interactions of particles and holes through such a formalism, we are able to move beyond the single particle framework to describe electron hole pairs as *excitons*: a type of quasiparticle, the behaviour of which is fundamental to understand processes in solar cell materials, etc.

The concept of a quasiparticle can be useful to understand the connection of a many-body system to the independent particle picture that lies behind the widespread use of orbitals, or single electron wavefunctions, to describe electronic shell structure. Screening by the electronic medium of whatever material environment a single particle is in – the background electron density – decreases the effect of the usual Coulomb interaction. Within this picture we can retain the physical meaning of the independent particle picture, by thinking of our individual particles as screened quasiparticles.

Many-body techniques can therefore be used for the calculation of more accurate electronic structures than are obtainable with DFT, but they are generally unnecessary for finite systems for which other complications, in particular structural variation and imperfection, tend to be more important. They are mentioned here only to leave you with the knowledge that the limitations of DFT are well known, and can be improved on as and when needed.

Bibliography

[1] S. J. Clark and J. Robertson. Screened exchange density functional applied to solids. *Physical Review B*, 82:085208, 2010.

[2] J. Akola, M. Walter, R. L. Whetten, H. Hakkinen, and H. Groenbeck. On the Structure of Thiolate-Protected Au_{25}. *Journal of the American Chemical Society*, 130:3756–3757, 2010.

CHAPTER 5

Real Metal Clusters

In Chapter 3, we considered the different potentials predicted for simple alkali metals (found in group 1 of the periodic table), and the influence of those changing potentials on the resulting energy eigenvalues of the confined valence electrons. We also considered the influence of atomistic (lattice) structure, in comparison to the jellium model and its assumption of a uniform background.

Experimental data for different elemental metals contains a number of complicating factors, which we are now in a position to unpack. These include:

- The mass, or size of the ionic core of the metal.

- The charge of the ionic core – or equivalently, the valency of the element.

- The orbital symmetry (s, p, ...) of the valence electrons.

- The lattice structure (or lack thereof) of the arrangement of atoms.

- The overall shape of the cluster – due to both electronic distortions and atomistic bonding constraints.

In this chapter, we will first examine experimental data on clusters of different metals and valency, before looking at how these complexities affect the idea of metal clusters as being 'superatomic'.

5.1 MONOVALENT METALS

Monovalent metals – in particular, those that appear in the first group of the periodic table, with a single valence s-electron per atom – have been very well studied. Indeed, sodium clusters are often considered the prototypical example of electronic shell structure in clusters, due to the early and influential work of Knight and others, which first demonstrated the size-dependence of cluster abundance in mass spectra [1], discussed already in Chapters 2 and 3.

There are clear similarities in the abundance plots of other group one metal clusters. For example, both sodium and potassium clusters, presented in Fig. 5.1, agree clearly on the electronic magic numbers expected from

DOI: 10.1201/b23295-5

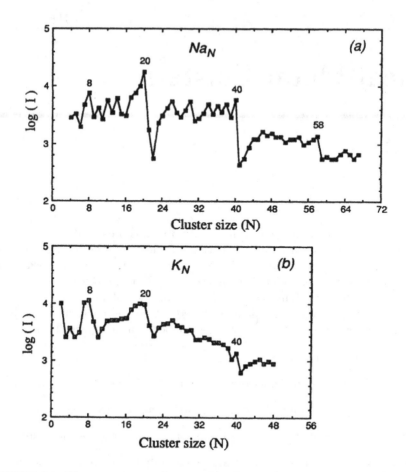

FIGURE 5.1: The experimental mass abundance plots of the group 1 metal clusters, of sodium and potassium, are compared. Figures reproduced with permission [1, 2].

the simple harmonic oscillator solutions: there are clear peaks at 8, 20, and 40 electrons, corresponding to the $1S^2, 1P^6$, $1S^2, 1P^6, 1D^{10}, 2S^2$, and $1S^2, 1P^6, 1D^{10}, 2S^2, 1F^{14}, 2P^6$ configurations, respectively. However, there are differences. For example, the even-odd oscillation in stability that is so pronounced for sodium up until $N = 40$ is not evident in the potassium spectrum, and the relative stabilities of the heavier clusters are reduced systematically in the case of potassium, compared to sodium. The even-odd oscillation is due to the significantly larger Fermi energy of sodium compared to potassium: it is the Fermi energy that determines the relevant energy scale of the electronic structure. On the other hand, the systematic decrease in relative size of the potassium clusters is more likely due to the increased mass of potassium, compared to sodium [2].

It is not only the group 1 metals that have a single valence electron, of course. The coinage metals, in group 11, are so called due to their stability in elemental form, which is in part due to their $d^{10}s^1$ electronic structure. The thermodynamic stability associated with closure of the d-shell means that their single valence electrons have s-character. They should, therefore, behave in similar ways to the group 1 metals. In Fig. 5.2, the relative abundances of charged silver and copper cluster ions are compared. These clusters are either positively or negatively charged (this is useful experimentally, to enable size-selection), and therefore demonstrate enhanced stability at $N = 9, 21, 41$, in the case of the positively chaged clusters, or at $N = 7, 19, 39$, for the negatively charged clusters – both corresponding to the atomic sizes at which the usual electronic shell closings are satisfied.

5.2 DIVALENT METALS

With two valence electrons per atom, one might worry that the divalent metals would exhibit less clear electronic shell structure. There are indeed cases where this is true, such as is shown for barium, a heavy metal in group 2, in Fig. 5.3. The numbers observed in this case – $N = 13, 19, 23, 26, \ldots$ – do not correspond to electronic shell closings. In fact, they correspond to geometric shell closings, expected for atoms that prefer to pack in icosahedral symmetry: the smallest shell closing consists of a single atom surrounded by 12 equally spaced nearest neighbours, while 19 atoms is the number obtained when two atoms are each surrounded by 12 nearest neighbours, of which they are each one, and then share 5. The second perfect icosahedron is obtained at 55 atoms, after placing additional atoms on each of the 20 faces of the 13 atom icosahedron, as well as directly over each of the 13 atoms themselves.

It is possible to interpret this result as a consequence of the reduced metallicity of barium at small cluster sizes: unlike for monovalent metals the valence band is fully occupied, while the conduction band is empty, and therefore a band gap will exist at very small sizes.

In contrast, the abundance plots for zinc and cadmium clusters, for which the atomic configuration is $d^{10}s^2$, show a very different behaviour, as seen in Fig. 5.4. Consistent peaks appear at $N = 10, 20, 28, 35, \ldots$, and as these clusters are charged, the presence of even numbers demonstrates immediately that there is a stability associated with even numbers of atoms that is competing with the electronic shell closings. However, the corresponding electronic numbers (allowing for two s-electrons per atom), $N_e = 19, 39, 55, 69$, are as close as they can be to shell closings, given that the overall number of electrons is constrained to be odd. Thus, we still see the importance of this electronic shell structure.

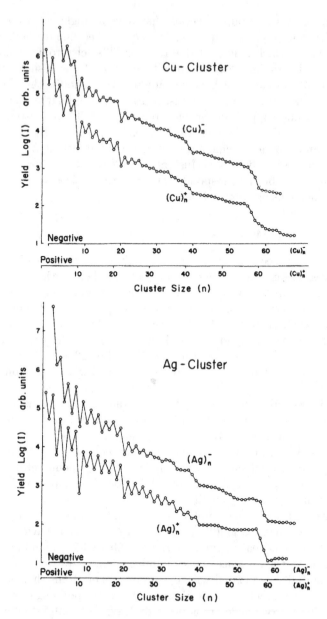

FIGURE 5.2: The experimental mass abundance plots of the charged group 11 metal clusters, of copper and silver, are compared. Figures reproduced with permission [3].

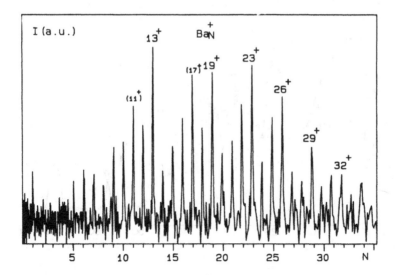

FIGURE 5.3: The experimental mass abundance of the group 2 metal clusters, of barium, is presented. Figure reproduced with permission [4].

5.3 TRIVALENT METALS

5.3.1 Trivalent Metals with s^2p^1 Configurations

The simplest trivalent metals are those with electronic configuration s^2p^1, which occur in group 13, and include aluminium, gallium, and indium. Of these, the lightest congener has the most simple metallic behaviour, and for this reason, aluminium clusters have been studied the most extensively. To satisfy the usual electronic shell structure, we need to find electronic magic numbers that are (approximately) divisible by three. A classic example is the 40 electron shell closing: this is obtained from an icosahedral structure of 13 trivalent atoms, with the addition of a single additional electron. For this reason, Al_{13} and its anionic equivalent have been extensively studied.

Early analysis of the photodissociation energies of aluminium clusters demonstrated clear maxima in the stability of the positively charged cluster ions with 7 and 13 atoms [6]. When positively charged, these cluster sizes correspond to 20 and 38 electrons respectively; the first of these is a shell closing (of the $2S$ shell) which the second is close to the shell closing expected at 40 electrons, which is however inaccessable to cationic clusters of aluminium due to the 3 valence electrons per atom. This provides some evidence that proximity to a shell closing may confer some stability, however, the geometric stability of the icosahedral 13 atom cluster is likely also a factor.

While the 40 electron shell closing in Al_{13}^- makes it a closed shell atom, it is important to recognise that the shell closings only tell us when we have an

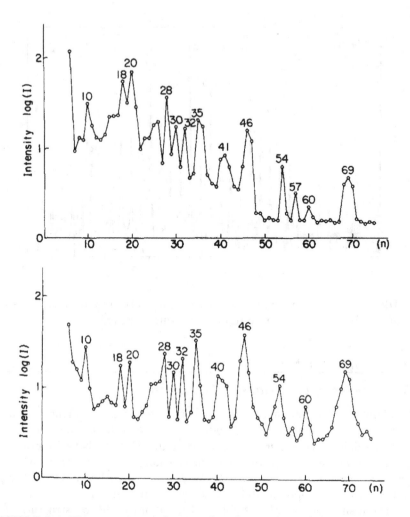

FIGURE 5.4: The experimental mass abundance plots of the group 12 metal clusters, of zinc and cadmium, are compared. Figure reproduced with permission [5].

atom that is analogous to a noble gas atom. A superatom with 39 electrons is still a superatom, if it exhibits the same shell structure – it merely has an electronic vacancy that would like to be filled, in exactly the same way as for the halogens, fluorine and chlorine, for example. This observation led to the demonstration in 2004 that Al_{13} behaves as a superatom [7].

The generality of this behaviour for aluminium clusters has been demonstrated in a range of works. In particular, the demonstration that the Al_{13} motif could be extended to a series of halogen mimics described by the formula $Al_{13}I_{2n}^-$, where n is an even number, and to clusters of $Al_{14}I_{2n+1}^-$, with n

$$\text{Al}_{14}\text{I}_3^-$$

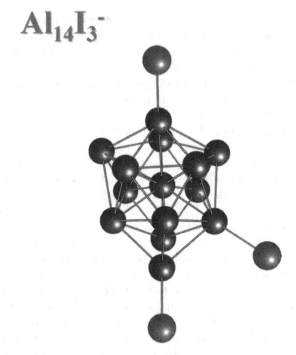

FIGURE 5.5: The geometric structure of the $\text{Al}_{14}\text{I}_3^-$ cluster found to be an alkaline earth atom mimic as a superatom. The three iodine atoms are external to the aluminium core, which largely retains the structure of the Al_{13} cluster. Reproduced with permission from [8]; Copyright 2011, American Chemical Society.

a odd number, prodived a starting point for the design of such entities [8]. Fig. 5.5 includes an image of the structure of an $\text{Al}_{14}\text{I}_3^-$ cluster whic is a mimic of a divalent alkaline earth atom, for example, representing a transition to the opposite side of the periodic table! It is also notable that the core structure of the Al_{13} moiety remains evident in this aluminium iodide cluster.

The study of aluminium clusters in particular, and their predictable bond formation with other elements in accordance with the principles of chemical bonding, has led to the realisation that these clusters can be used as atomic mimics, not only by exploiting their stability, but instead by understanding and controlling their electronic reactivity. One way of summarising the potential of these clusters has been to propose the idea of a three-dimensional periodic table, in which cluster size is used to control shell structure, rather than nuclear charge [10].

For all the extensive work on aluminium clusters, it is not the only trivalent metal, although it is the most metallic element in the p-block of the periodic table, where the elements that are metallic are sometimes referred to as the *poor metals*, due to the introduction of an increasing covalent character to their interatomic bonding, induced by the presence of p-electrons, and that leads to the non-metallic character of the elements to the right of the periodic table. The element above aluminium is boron, which as a non-metallic element we will exclude from the current discussion; the element below aluminium in this group is however gallium, a metal that has some number of idiosyncracies, such as a room temperature melting point, but is nonetheless a metal with some similarities to aluminium.

While aluminium and gallium have different lattice structures in the bulk, their cluster structures are rather similar, in the size range of tens of atoms that is most relevant to our discussion of superatoms [11]. Thermodynamic studies of gallium clusters have demonstrated that the clusters have much higher thermal stability than the bulk, again demonstrating that the clusters behave more like aluminium [12, 13]. While, as for aluminium, the trivalent character of gallium means that not all magic numbers are observable for the clusters as a function of size, the case of Ga_{13} is instructive. Simulations of the melting transition of this cluster as a function of charge state – in the neutral, anionic, and cationic forms – have shown that the shell closing due to the addition of a single electron to create the 40 electron system results in a significant improvement in thermal stability, as the melting temperature is increased by $\sim$200 K [9]. Even more interestingly, at finite temperature (as simulated using molecular dynamics simulations that are based on a DFT calculation of electronic structure), the enhanced stability is shared to an extent by the (neutral) 39 electron system when compared to the 38 electron (positively charged) cluster: there is a $\sim$200 K difference in this case too, attributable to the fact that the 39 electron cluster has already occupied the highly symmetric superatomic orbital that closes at 40 electrons, and the high symmetry of this orbital stabilses the solid cluster. In Fig. 5.6 the average structure of Ga_{13} is shown, at both low temperature, where none of the atomic positions change much, and at very high temperature just before the melting transition, where all the surface atoms can be seen to interchange, but while retaining the high symmetry of the cluster.

The high symmetry of the Ga_{13} cluster is shown in Fig. 5.7 to relate to the high symmetry of the superatomic electronic shell populated by the highest energy electron. For both the neutral and anionic clusters, with 39 and 40 electrons respectively, this orbital has similar symmetry character, while the 38 electron system has considerably reduced symmetry, both when considering electron density and the atomic positions in the cluster. To the left of each molecular orbital image in Fig. 5.7, the DOS is shown, as projected onto spherical harmonic functions to reveal the symmetry progression of the states. The symmetry of each superatomic shell reflects the symmetry of the cluster; for example, the cationic cluster which is the least spherical, shows the clearest splitting of the $1P$ superatomic state in the DOS.

5.3.2 Trivalent Metals with s^2d^1 Configurations

It is not only the p-block metals that are trivalent. Amongst the transition metals, those in group 3 – scandium, yttrium, and lanthanum – have a s^2d^1

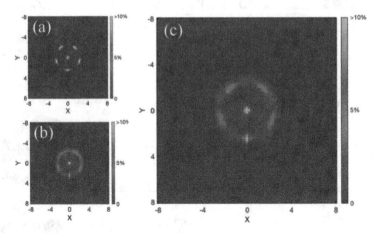

FIGURE 5.6: The averaged geometric structure of the Ga_{13} cluster, projected onto a plane at different temperatures. (a) the lowest simulated temperature ($\sim$200 K); (b) 1000 K. Panel (c) gives an overlay of the two structures. Reproduced from Ref. [9] with permission from the PCCP Owner Societies.

configuration. Experimental studies of relevance have shown that these metals, doped into an alkali metal cluster that is superatomic in nature, can contribute all their valence electrons into the electronic shell structure, but experimental evidence for the pure clusters of these metals is not so clear. However, recent DFT calculations on these cluster, focusing on the clusters of 7 atoms of different charge states, and allowing for the full range of spin states to be explored, have shown that there is a systematic progression of electronic shells to be considered, which include the d-electrons. These clusters all have the same atomistic structure: a ring of five atoms, with a capping atom above, and one below the ring, is the favoured geometry (the importance of this will be evident in the discussion below). However, there are considerable differences between these clusters and the more traditionally studied superatoms [14].

In Fig. 5.8, the angular momentum characters of the molecular orbitals of the group 3 clusters are given. In accordance with the superatomic concept, the first 8 electrons occupy a $1S^2 1P^6$ configuration, in all cases. However, the following series of shells – $1D, 2S$, and $1F$ – follow in an irregular fashion for the three different elements. This can be partly understood by the fact that these clusters are magnetic; they all have a number of unpaired electrons, resulting in a particular spin state (the state shown in this figure is the one that has been calculated to be of lowest energy in this case, using DFT (with a PBE0 functional description of the many-body electron-electron interactions). More insight is possible if we count electrons a little more closely: there are 2 s-electrons contributed by each atom in the cluster, and thus we might expect the first 14 electrons to contribute to standard superatomic shells, as

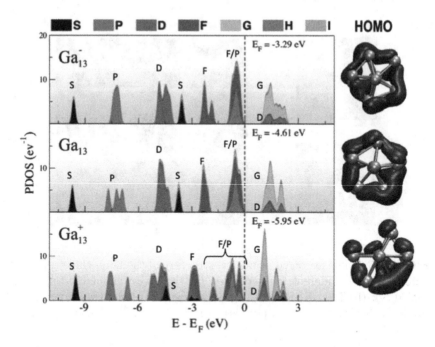

FIGURE 5.7: The PDOS of the Ga$_{13}$ cluster, for each charge state, with a visualisation of the electron density of the highest energy superatomic orbital presented at right. These are snapshots of the finite temperature cluster structure, taken at temperatures ∼300 K below the melting temperature. Reproduced from Ref. [9] with permission from the PCCP Owner Societies.

we know that s-electrons do. However, the 7 highest energy molecular orbitals in each case must be due to the atomic d-electron contributed by each atom. Notably, these appear in all respects to be equally delocalised over the whole cluster as the s-electron based molecular orbitals, and are equally assignable to particular spherical harmonic symmetries in most cases. This shows that the d-electrons are capable of contributing to superatomic shell structure.

There is one exception to the statement that these molecular orbitals are assignable to spherical harmonic symmetries, and that is the case of the case of the orbital labelled with an X, that appears as the highest energy orbital, for example, for the Sc$_7$ cluster. This has a five-fold symmetry that reflects the underlying symmetry of the cluster, and thus demonstrates that molecular orbitals formed from d-orbitals will be more readily influenced – energetically stabilised or destabilised – by the underlying symmetry of the cluster, due to the atomic positions, than molecular orbitals constructed from s-electrons. This realisation helps us to interpret the meaning of the mixture of $1D, 2S$, and $1F$ states, that do not seem to follow the usual superatomic tendency for

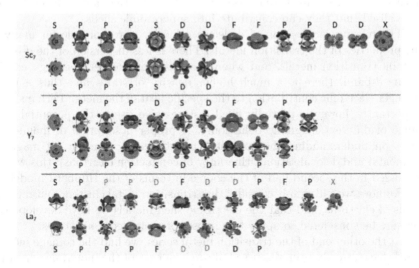

FIGURE 5.8: The molecular orbitals of the Sc_7, Y_7, and La_7 clusters are visualised in energetic order from left to right. The assigned angular momentum labels are given for each orbital, based on projection of each orbital onto spherical harmonics; the label is in black if the projection value is in excess of 0.36, and in grey if the value is greater than 0.2 for a given spherical harmonic. The top row of orbital images is those of majority spin, while the second row is for those of minority spin. Reproduced from Ref. [14] with permission from the PCCP Owner Societies.

shell closing: some of the $1D$ and $1F$ molecular orbitals are oriented along the appropriate axes to be compatible with the cluster symmetry, and thus are energetically favoured (and occupied), while those that are oriented in a disfavoured way relative to the cluster structure are energetically destabilised.

In summary, we learn from this study that transition metal elements may also be superatomic, but that due to (a) the ability of these metals to support high spin states, and (b) the ability of d-orbitals to break degeneracy due to local symmetry considerations, the interpretation of these clusters in terms of the superatomic model may be non-trivial.

5.4 TRANSITION METALS

The recognition that higher angular momentum wavefunctions have reduced radial extent, has meant that for a long time the transition metal elements (those that have an open d-shell in some elemental or compound form) were excluded from consideration within the superatomic model. However, as we have just seen, group 3 elements have d-electrons that are sufficiently well delocalised that they can contribute into superatomic shells.

The relative localisation of d-electron density is responsible for many of the properties of the transition metals. There is less dispersion of the d-bands in bulk transition metals, and when the states at the Fermi energy consist of the d-band, there is a much higher density of states and thus a much stronger electronic contribution to the specific heat of the metal. Perhaps most importantly, however, partially filled d-states give rise to the potential for a range of different magnetic behaviours. A proper description of magnetism relies on understanding the competition between localisation (of magnetic moments) and delocalisation (itinerancy of conduction electrons): this will be discussed in the second half of the course in terms of the Hubbard model.

For now, we will simply consider that transition metal clusters contain both kinds of electrons, and that the electronic shell models we have developed so far may be considered to apply to a subset of their electronic states.

At the other end of the transition metal series, we find the coinage metals, which have already been discussed above as a case of the monovalent metals, where the single valence s-electron participates in superatomic shells, due to the closed and therefore inert nature of the atomic d-shell.

The case of nickel clusters, therefore, one group to the left of copper in the series, is particularly intriguing. Nickel has a $3d^8 4s^2$ electronic configuration as an atom, but under certain conditions this can adjust to be $3d^9 4s^1$ or even $3d^{10} 4s^0$ due to the energetic proximity of the $3d$ and $4s$ states. In the small clusters, a systematic preference has been found for a $3d^9 4s^1$ configuration, allowing for 9 electrons to remain localised, producing magnetic moments that control the overall magnetism of the cluster. The superatomic shell structure is then formed from the relatively delocalised s-electrons, with an electron counting rule of one electron per atom, just as for gold [15].

However, the case of Ni_{10}^+ is a particularly intriguing exception, as shown in Fig. 5.9. Due to the energetic stabilisation of the closed superatomic $1P$ state, which is that of the highest energy electron, Ni_{10}^+ prefers to adjust its total electronic configuration to obtain a $1S^2 1P^6$ configuration, with 8 superatomic electrons. The available number of electrons is 9: one per atom, but with one s-electron removed due to the positive charge (the s-electrons being less tightly bound than the d-electrons). Thus the preferred cluster structure has a central atom in the $3d^{10}$ configuration, surrounded by 8 atoms in the $3d^9 4s^1$, in order to maximise its electronic stability.

This intriguing example of the coexistence and competition between localised and delocalised states in transition metal clusters provides a first demonstration that the superatomic model may be used to interpret the magnetic properties of small nanoclusters.

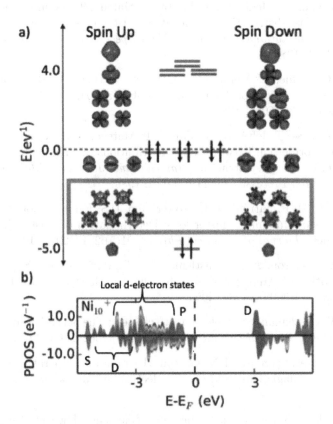

FIGURE 5.9: The electronic eigenvalues and corresponding orbitals are presented (a), and the corresponding density of states with projected symmetry assignments (b), are presented for the Ni_{10}^+ clusters. Reproduced from Ref. [15] with permission.

Bibliography

[1] W. D. Knight, K. Clemenger, W. A. De Heer, W. A. Saunders, M. Y. Chou, and M. L. Cohen. Electronic shell structure and abundances of sodium clusters. *Physical Review Letters*, 52(24):2141–2143, 1984.

[2] W. A. de Heer. The physics of simple metal clusters: experimental aspects and simple models. *Reviews of Modern Physics*, 65(3):611–676, 1993.

[3] I. Katakuse, T. Ichihara, Y. Fujita, T. Matsuo, T. Sakurai, and H. Matsuda. Mass distributions of negative cluster ions of copper, silver, and gold. *International Journal of Mass Spectrometry and Ion Processes*, 74:33–41, 1986.

[4] D. R Rayane, P. Melinon, B. Cabaud, A. Hoareau, B. Tribollet, and M. Broyeri. Close-packing structure of small barium clusters. *Physical Review A*, 39: 6056, 1989.

[5] I. Katakuse, T. Ichihara, Y. Fujita, T. Matsuo, T. Sakurai, and H. Matsuda. Correlation between mass distributions of zinc, cadmium clusters and electronic shell structure. *International Journal of Mass Spectrometry and Ion Processes*, 69(1):109–114, 1986.

[6] U. Ray, M. F. Jarrold, J. E. Bower, and J. S. Kraus. Photodissociation kinetics of aluminum cluster ions: Determination of cluster dissociation energies. *Journal of Chemical Physics*, 91(5):2912–2921, 1989.

[7] D. E. Bergeron, A. W. Castleman Jr., T. Morisato, and S. N. Khanna. Formation of $Al_{13}I^-$: Evidence for the superhalogen character of Al_{13}. *Science*, 84(2009):84–88, 2004.

[8] A. W. Castleman, From elements to clusters: The periodic table revisited. *Journal of Physical Chemistry Letters*, 2(9):1062–1069, 2011.

[9] K. G. Steenbergen and N. Gaston. Ultra stable superatomic structure of doubly magic Ga_{13} and $Ga_{13}Li$ electrolyte. *Nanoscale*, 12(1):289–295, 2020.

[10] Puru Jena. Beyond the periodic table of elements: The role of superatoms. *Journal of Physical Chemistry Letters*, 4(9):1432–1442, 2013.

[11] K. G. Steenbergen, D. Schebarchov, and N. Gaston. Electronic effects on the melting of small gallium clusters. *Journal of Chemical Physics*, 137(14):144–307, 2012.

[12] K. G. Steenbergen and N. Gaston. A two-dimensional liquid structure explains the elevated melting temperatures of Gallium nanoclusters. *Nano Letters*, 16(1):21–26, 2016.

[13] G. Breaux, R. Benirschke, T. Sugai, B. Kinnear, and M. F. Jarrold. Hot and solid Gallium clusters: Too small to melt. *Physical Review Letters*, 91(21):19–22, 2003.

[14] J. T. A. Gilmour and N. Gaston. On the involvement of d-electrons in superatomic shells: The group 3 and 4 transition metals. *Physical Chemistry Chemical Physics*, 21(15):8035–8045, 2019.

[15] J. T. A. Gilmour, L. Hammerschmidt, J. Schacht, and N. Gaston. Superatomic states in nickel clusters: Revising the prospects for transition metal based superatoms. *Journal of Chemical Physics*, 147(15), 2017.

CHAPTER 6

Non-metal Superatoms

The simple picture of a superatom based on electronic structure formed from the valence electrons of a cluster of metal atoms is a powerful one. It unifies many of the experimental observations made on these systems to date, as seen in the previous chapter, and we will see in subsequent chapters how this model provides predictability of how the properties of superatoms can be controllably tuned for materials design.

However, a discussion of superatoms needs to acknowledge the existence of competing ideas relating to superatomicity in the literature, some of which do not rely on the predictable behaviour of valence electrons in metal clusters at all. In contrast, the concept of a superhalogen, or a superalkali, was coined in order to describe the behaviour of molecules that can be designed to have key properties that exceed those of any natural atoms. In particular, halogen atoms have the highest electron affinities of any atoms in the periodic table, and thus any species found to have a higher electron affinity than a natural halogen atom may be called a superhalogen. Similarly, alkali metals in group one of the periodic table have the greatest ability to lose their valence electrons, and thus the lowest ionisation potentials: this meant that any species found to have a lower ionisation potential than a natural alkali metal may be called a superalkali.

While these definitions are functional, they say nothing about the electronic structure of the species itself. In this chapter, we will therefore explore the nature of a few well-studied examples, in order to summarise the ways in which these molecules do and do not conform to the general ideas about superatoms that we have outlined so far in this book.

6.1 SUPERHALOGENS

The concept of a 'superhalogen' was demonstrated in a pair of 1981 papers by Gutsev and Boldyrev [1, 2], in which they used then state of the art electronic structure theory to calculate the electron affinities of the molecules $BeCl_3$, BCl_4, $MgCl_3$, $AlCl_4$, $SiCl_5$ and PCl_6, and similarly constructed fluorine and oxygen containing molecules. These all have in common that they are

DOI: 10.1201/b23295-6

composed of a small central atom from the first or second row of the periodic table, surrounded by as many halogens – such as chlorine – as can be made to fit. In essence, this is systematically one more halogen than the number of valence electrons of the central atom: so for example, in the case of Al which has a s^2p^1 valence configuration, the first 3 chlorine atoms can be thought of as bound directly through a bond that also requires one of these valence electrons from the central atom. The additional halogen atom in each molecule is therefore bound cooperatively – which is to say, the strong polarisation of electron density between the central atom and the halogen ligands is sufficiently symmetric that if an addiitonal halogen atom fits on the surface of this sphere, the bonding will be shared equally between all halogens that can fit around the metal centre.

The fact that there is an additional halogen in each case, however, means that each of these molecules, in neutral form, is severely electron deficient. This means that the ability of each of these molecules to attract an electron from elsewhere, as measured by the calculated electron affinity, is higher than for an individual halogen on its own – justifying the use of the term superhalogen.

Since the original work, this same structural motif – a central stabilising metal atom that can carry multiple halogens and concentrate their properties into something we call a superhalogen – has been widely employed, in particular through the use of fluorine, which is itself even more electronegative (electron accepting) than chlorine.

This description of the design of these molecules – as a way of concentrating the electronegativity of individual halogens in space in order to create a greater electron affinity of the molecule – is simplistic, but a reasonable description of the idea nonetheless. So what, if anything, does it have in common with the jellium-model like description of clusters of metal atoms that we have used as the basis of our understanding of superatoms up till now?

Some of the key principles have been elucidated through the study of the importance of symmetry in determining the electron affinities of these molecules, as well as through analysis of where the additional electron in any of the negative forms of these superhalogens actually ends up [3]. By testing the effect of combining different halogens (e.g., both fluorine and chlorine) in the same molecule, they were able to show that the high symmetry only achievable when there is a single type of halogen atoms strongly enhances the electron affinity. This importance of spherical symmetry is reminiscent of what we have seen in the analysis of metal cluster based superatoms already.

The key difference that one might expect to see in a molecule of this kind, in contrast to a metallic cluster, is that it might not have the same delocalisation of electron density over the whole object, in order to enable us to think of it as a single electronic unit. This question is most easily addressed by analysis of the additional electron that is added to the neutral molecule in order to form the anion: is it localised in any particular part of the molecule, or is it delocalised over the whole structure?

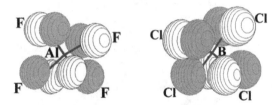

FIGURE 6.1: The symmetric nature of the highest occupied molecular orbital (HOMO) in the superhalogen anions AlF_4 and $AlCl_4$ is shown. Figure used with permission [3]. Copyright 2008, American Chemical Society.

As seen in Fig. 6.1, the highest occupied molecular orbital (HOMO) in the highly symmetric superhalogen anions, which indicates the location of the accepted electron, is equally distributed across all four halogens, as seen by the p-type electron density centred on the presence of each halogen ligand. This is further evidence that thinking of these bonds as cooperative ionic bonds on the surface of a sphere around the central atom is essentially correct, and provides further analogy to the spherical shell structure we are used to seeing in the case of metal cluster superatoms.

6.2 SUPERALKALIS

In 1982, Gutsev and Boldyrev [4] followed up their description of superhalogens with a study of small molecules formed from clusters of alkali metal atoms Li, Na, and Cs, bound togther with a more electronegative atom (such as oxygen or nitrogen) so as to form a radical species – one in which, as in the case of the superhalogens above, there is a odd number of electrons. The ionisation potentials of these species are in all cases less than those of the original alkali metals, meaning that the outermost electron in the radical is very weakly bound, and thus the cation is even more easily formed. This results in the description of such species as superalkalis.

As in the case of the superhalogens, the molecular character of these superalkalis makes them rather different than the metal cluster based superatoms that we have already discussed. In particular, the superalkalis are not in general spherical species, with some of them even adopting linear geometries. On the other hand, they do contain multiple metal atoms, as a rule, and therefore have some metal cluster character.

In Fig. 9.1, the electronic character of a recently described class of superalkalis based on a N_4Mg_6M structural motif is elucidated, for the cases where

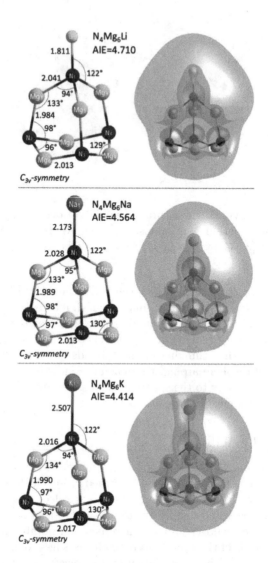

FIGURE 6.2: The ground states of the non-charged N_4Mg_6Li, N_4Mg_6Na, and N_4Mg_6K species and their HOMOs (plotted with a fraction of electron density equal to 0.8). Adiabatic ionisation energies (AIE, in eV) and bond lengths (in Å) are also provided. Reproduced from [5], with permission.

M = Li, Na, and K. The HOMOs of the neutral species are clearly delocalised over the entire molecule, meaning that this easily donates electron is accessible through interactions with the whole molecule, rather than through any single constituent. This is confirmed through analysis of the interactions of these superalkalis with the particularly stable CO_2 molecule; rather than the

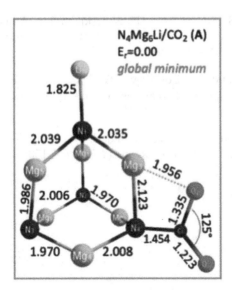

FIGURE 6.3: The lowest energy structure of the N_4Mg_6Li – CO_2 complex is shown, with bond lengths (in Å) and the strong effect on the geometry of the linear geometry of the CO_2 molecule shown. Reproduced from [5], with permission.

preferred interaction being through the most electropositive metal atom, the CO_2 interacts most strongly when located in close proximity to one of the three equivalent N atoms, as shown in Fig. 9.2.

In summary, the long history of analysis of these superhalogen and superalkali species, motivated by their ability for strong and tunable electronic interactions, can teach us something about the prospects for the more general class of metal cluster based superatoms. While they do not conform to the most general picture of electronic shell structure provided by the Jellium model, they nonetheless challenge the distinction between metal atom based, delocalised electron density systems, and molecular systems that might more naively be thought to comprise electron density localised in covalent bonds.

Bibliography

[1] G. L. Gutsev and A.I. Boldyrev. DVM-XQ calculations on the ionization potentials of MX_{k+1}^- complex anions and the electron affinities of MX_{k+1} "Superhalogens". *Chemical Physics*, 56:277–283, 1981.

[2] G. L. Gutsev and A. I. Boldyrev. DVM $X\alpha$ calculations on the electronic structure of complex chlorine anions. *Chemical Physics Letters*, 84:352–355, 1981.

[3] C. Sikorska, S. Smuczyńska, P. Skurski, and I. Anusiewicz. BX_4^- and AlX_4^- superhalogen anions (X = F, Cl, Br): An ab initio study. *Inorganic Chemistry*, 47:7348–7354, 2008.

[4] G. L. Gutsev and A.I. Boldyrev. DVM Xα calculations on the electronic structure of "Superalkali" cations. *Chemical Physics Letters*, 84:352–355, 1981.

[5] C. Sikorska and N. Gaston. N_4Mg_6M (M = Li, Na, K) superalkalis for CO_2 activation. *Journal of Chemical Physics*, 153:144301, 2020.

CHAPTER 7

Ligand Protected Metal Clusters

In Chapter 5, we considered the ways in which different metal elements, with their different valency and size, lead to different perturbations of the internal electronic structure of a cluster away from the simple jellium model. However, metal clusters do not generally exist in pure metallic form in the gas phase. In addition to the internal structure that arises purely from the nature of the metal atoms themselves, the surface environment of the cluster is also crucial to understand, when considering its effect on the confinement potential, as well as the way the clusters may interact with other objects, such as other clusters.

We will therefore examine the types of metal clusters which truly reinvented the relevance of the superatomic concept, by making it applicable to clusters outside of vacuum chambers, and that can exist in solution. These clusters are in essence the same as those we have considered so far, but with one crucial difference: the surface of the cluster is protected from oxidation, or other forms of chemical reaction, by a protective shell of small molecules that attach to the surface and make the whole nanoparticle soluble in a suitable organic or aqueous solvent.

A ligand is a molecule that bonds to a metal atom. The bond can range between ionic and covalent in character, which influences the distribution of electronic charge at the surface of the cluster. We refer to metal clusters stabilised in this way as ligand protected. They may also be referred to as metalloid: while the surface of the cluster consists of atoms that have formed bonds with ligands, which may effectively localise the valence electrons and make that atom less metallic, there remain sufficient metal atoms that are

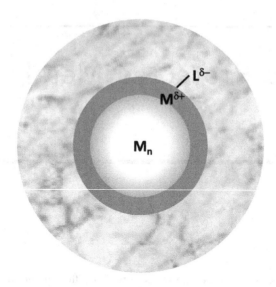

FIGURE 7.1: A ligand-protected cluster has a number (n) of metal atoms (M) at the core that retain the valence (conduction) electrons, while metal atoms at the surface may instead be partially oxidised, and have a partial positive charge. These surface metal atoms $M^{\delta+}$ will each be bonded to a ligand, $L^{\delta-}$, forming a ligand shell that protects the metal cluster from further oxidation or chemical reaction, but that can also be mobile and porous, depending on its composition.

not bonded to ligands, to retain delocalised electronic states and metallic character.

The metals most usually seen in these types of clusters are those that are most readily stabilised in this way – for this reason, particularly reactive metals such as sodium are not as easily synthesised in ligand-protected cluster form as metals such as gold, which as a coinage, or noble metal, is relatively inert. Copper and silver, and aluminium and gallium, have also been widely studied, and will be discussed in the rest of this chapter.

7.1 GOLD THIOLATE CLUSTERS

The application of the superatomic model to ligand protected complexes, and the subsequent revision of the model to account for ligands, was clearly demonstrated in work published in 2008 by Häkkinen et al. [1]. Gold is well known to like forming bonds to sulfur, so it is no surprise that thiolate ligands are successful in stabilising gold clusters (a thiolate molecule is a carbon based molecule that contains a sulfur atom that is bound to a hydrogen atom in its neutral form, but that can replace the bond to the hydrogen atom to a bond

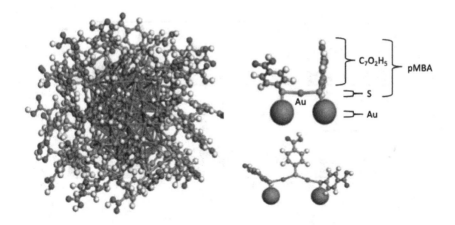

FIGURE 7.2: The structure of a 102 atom gold cluster, protected with MBA ligands. Left: the full structure, showing the extent of the ligand shell around the cluster; Right: two different local assemblies of the ligands, demonstrating the abstraction of gold atoms into the ligand shell. Figure adapted with permission from Ref. [1], Copyright (2008) National Academy of Sciences, U.S.A.

to gold). In so far, the bonding between gold and sulfur is simple, but in when many metal atoms are involved, cooperativity can lead to unexpected structures. In this case, it was demonstrated that the sulfur ligands can pair up on either side of an oxidised gold atom, forming a kind of molecular 'staple' that holds the aggregated ligand structure to the underlying metal core. Alternatively, assemblies of three thiolate molecules, with two gold atoms removed from the core of the cluster, can be seen to form. The structure of such a cluster protected by *para*-mercaptobenzoic acid, $SC_7O_2H_5$ ligands, abbreviated as MBA is shown in Fig. 7.2 and its density of states is presented in Fig. 7.3.

The nature of the bonding is important for the understanding of these superatomic clusters as it affects the number of electrons that we consider participate in the electronic shell structure of the metal core. The ligand molecules, being primarily carbon-based, are composed of covalent bonds that localise electrons between atoms. These electrons, therefore, do not participate in the delocalised electron density that is confined within the cluster. This distinction in in part evident already in the DOS presented in Fig. 7.3.

While there are 102 gold atoms in the cluster, the gold atoms that are segregated from the cluster core by being covalently bound to the thiolate molecules have their valence s-electrons localised in these covalent bonds. Therefore, there remain only 79 gold atoms in the cluster core, and the PDOS of this isolated cluster is shown in the lower half of Fig. 7.3. Naturally, with 79 gold atoms, there are 79 s-electrons delocalised across the cluster core, and

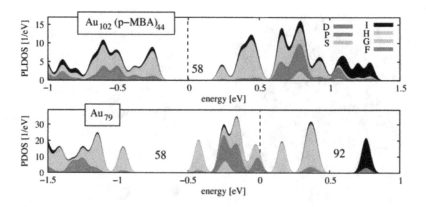

FIGURE 7.3: The density of states of a 102 atom gold cluster protected with MBA ligands. Top: the DOS of the ligand-protected cluster is shown. Bottom: the DOS of just the gold core is calculated, showing that the superatomic shell closing at 58 electrons is also apparent in the bare cluster. Figure adapted with permission from Ref. [1], Copyright (2008) National Academy of Sciences, U.S.A.

thus the projection of the electron density onto spherical harmonics shows the presence of a $1H$ electronic shell that is partially filled.

The PDOS of the full cluster, with ligands still attached, is shown in the upper half of Fig. 7.3 and has both similarities and differences. While the overall progression of states looks similar, the highest occupied energy level (separated from the unoccupied electron shells by the vertical dashed line at the Fermi energy, which is the zero of the energy axis) is the $1G$ shell that is fully occupied with a total of 58 electrons. What this tells us is that an additional 21 electrons are localised at the surface of the metal cluster, in the bonds that connect the gold cluster to the ligand shell.

The important point here, is that while the details of the gold-sulfur system are quite unique, this general picture of superatomic electron shell structure existing in a metallic core, and a ligand shell containing localised electron density that can modify the electron count of the core, provides a unified view of ligand-protected gold clusters as superatom complexes, that has been able to be extended to a wide range of other metals, and different ligand types. An outline of this theory starts with the idea that the familiar electronic magic numbers:

$$n^* = 2, 4, 10, 20, 40, \ldots \qquad (7.1)$$

are satisfied by ligand protected clusters, where

$$n^* = Nv_A - M - z \qquad (7.2)$$

with the number (N) of core metal atoms and the atomic valence, v_A), the number M of electron-localising (or electron-withdrawing) ligands (assuming

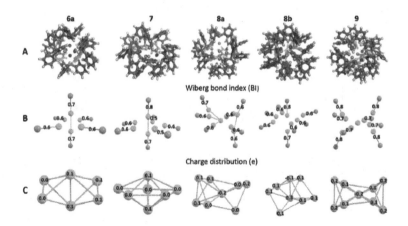

FIGURE 7.4: The structure of phosphine protected gold clusters, with 6 to 9 gold atoms. The strength of the gold-phosphine bonds is calculated and shown according to the Wiberg bond index, and the distribution of charge across the gold cluster core is also shown. Reproduced from Ref. [2] with permission from the PCCP Owner Societies.

here a withdrawal of one electron per ligand), and from the overall charge on the complex (z). The number of electrons per atom depends on the valency of the metal atom, as we have already seen; changing ligands can have similar consequences, as not all ligands withdraw (or localise) electron density to the same extent.

7.2 PHOSPHINE PROTECTED GOLD CLUSTERS

While sulfur based ligands withdraw electron density from the gold core, this is not necessarily the behaviour of all ligands, some of which can instead actually donate electron density to the superatomic core. This is the case for phosphine ligands, as phosphorus had five valence electrons, and forms three covalent bonds in a phosphine molecule, leaving a lone pair of electrons which can donate into the metal cluster. The structures of phosphine protected gold clusters of a range of sizes (number of gold atoms) are presented in Fig. 7.4, along with analysis of the bonding and charge distribution. The energy level analysis for these clusters has been calculated using DFT [2] and is presented in Fig. 7.5, firstly for the clusters with ligands removed (but for the same geometry of the gold atoms), and secondly for the clusters with the ligands included. This analysis shows that in each case, an additional two electrons, overall, are added into the delocalised electronic structure (the superatomic states) by the ligand shell.

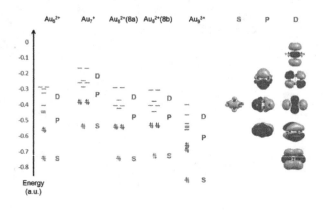

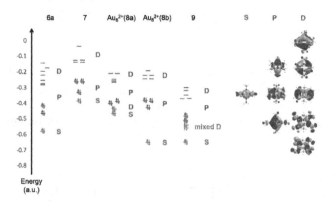

FIGURE 7.5: The electronic structure of phosphine protected gold clusters, with 6 to 9 gold atoms. Top: the energy levels are presented for the gold cluster core without ligands attached. Below: the energy levels are shown with ligands attached, demonstrating the persistence of the orbital symmetry and the contribution of electrons from the ligands to the cluster core. Reproduced from Ref. [2] with permission from the PCCP Owner Societies.

7.3 LIGAND PROTECTED ALUMINIUM AND GALLIUM CLUSTERS

Group 13 metals have been produced in a number of metalloid forms, with a range of different ligands used to protect them [3]. Much of the synthetic chemistry work has been conducted with the goal being to elucidate how the bulk metal is formed from individual atoms, with increasing size, treating the clusters as intermediates between the individual atoms and the bulk. However, over time, it has become apparent that the different structures formed at different sizes with the use of different ligands are often explainable only through the superatomic concept, and may then have very little in common with the structures of the bulk metals,

As the metals in group 13 have three valence electrons, it is relatively simple for the surface atoms in these clusters to bind covalently to a ligand and lose one electron to a localised bond, while still contributing two delocalised valence electrons to the global electronic structure.

Figure 7.6 is presented the structure of a cluster of 50 aluminium atoms, protected by 12 cyclopentadienyl (C_5H_5) ligands, based on DFT calculations [4]. The 12 aluminium atoms bound to the ligands contribute a total of 24 valence electrons, while the remaining 38 aluminium atoms contribute 3 electrons each, leading to a total of 138 superatomic electrons. In the shell structure models we have examined previously, this corresponds to the electron configuration $1S^2 1P^6 1D^{10} 2S^2 1F^{14} 2P^6 1G^{18} 2D^{10} 1H^{22} 3S^2 2F^{14} 3P^6 1I^{26}$. This is shown in Fig. 7.7.

The colour coded PDOS in Fig. 7.7, with weighted contributions of each spherical harmonic to each density functional theory derived orbital plotted in the DOS, demonstrates the very clear succession of electronic shells even up to $1I$ shell, which is fully occupied in this case, conferring a particular stability on this cluster.

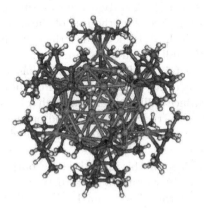

FIGURE 7.6: The highly spherical structure of a 50-atom aluminium cluster protected by organic Cp ligands. Reproduced with permission from Ref. [4].

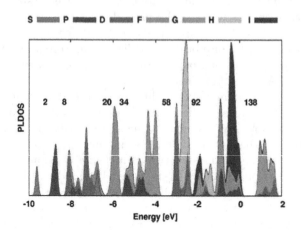

FIGURE 7.7: The density of states of a 50-atom aluminium cluster protected by organic Cp ligands, showing the 138 electron shell closing. Adapted with permission from Ref. [4].

7.3.1 The Role of the Ligand Shell

The element below aluminium in the periodic table, gallium, also forms a range of structures with ligands of this type. The structure of $Ga_23\{N(SiMe_3)_2\}_{11}$ has been calculated using DFT based on the experimental crystal structure [5] and is presented in Fig. 7.8. Here the role of the ligands has been further analysed by modifying the ligands included in the computation: the structure labelled 1a contains the full ligand structure from experimental data. The structure labelled 1b contains only NH_2 ligands, based on the assumption that the electronic effects of the ligand arise from the interaction between the nitrogen atom and a surface gallium atom, through a covalent bond, and that there should be no effect of truncating the ligand on the electronic structure. Finally, 1c depicts the same metal core with only hydrogen atoms used to mimic the effect of the ligands.

In Fig. 7.9, the validity of this hypothesis is outlined. Indeed, the exact same progression of superatomic shell structure is observed in all three cases, demonstrating that this cluster is a $1S^21P^61D^{10}2S^21F^{14}2P^61G^{18}$ superatom with 58 delocalised valence electrons. 22 of these electrons are contributed by the surface gallium atoms, which also contribute one electron each to a bond with the ligand atom, whether this is the original nitrogen based amine molecule or simply a surface hydrogen atom. 36 of these electrons are

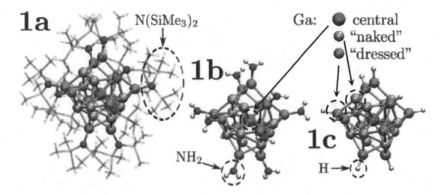

FIGURE 7.8: The structure of $Ga_{23}\{N(SiMe_3)_2\}_{11}$ is shown (1a), as well as model systems protected only by NH_3 ligands (1b) or H ligands (1c). Reproduced from Ref. [5] with permission from the PCCP Owner Societies.

contributed by the remaining gallium atoms at the core of the cluster, which remain unoxidised.

The symmetries of the calculated MOs are visualised in Fig. 7.9, above the PDOS of $Ga_{23}\{N(SiMe_3)_2\}_{11}$, and its simplified variants, for comparison. The nodal structure of these superatomic orbitals is clearly reminiscent of atomic orbitals, with 1, 2, 3, and 4 nodal planes visible upon inspection of the P, D, F, and G orbitals. This shows that the effect of the ligand shell on the superatomic electronic structure can be understood purely by counting covalent bonds to the gallium atoms at the surface.

In order to confirm the proposed distinction in the electronic configuration of the surface and internal gallium atoms, one option is to calculate the actual charge on each atom, to confirm the basis of the electron counting models above. The 'atoms in molecules' theory constructs local atomic charges (originally proposed by Bader, and therefore often referred to as Bader charges) by partitioning the charge density (ρ), by following the gradient at a particular point in space to the location of a charge density maximum centred at an atom. This defines so-called 'basins of attraction' of fixed points of the charge density. We can then define atomic charges as integrals over these basins, or Bader volumes, Ω_ρ. Each Bader volume contains a single electron density maximum and is separated from other volumes by a zero flux surface of the gradients of the electron density, $\nabla\rho(\mathbf{r})\dot{\mathbf{n}} = 0$. Here, $\mathbf{n}$ is the unit vector perpendicular to the dividing surface at any surface point on the volume.

The Bader charge analysis for the $Ga_23\{N(SiMe_3)_2\}_{11}$ is presented in Fig. 7.10 (in addition to analyses of the same cluster with simplified ligand structures). This analysis confirms unambiguously that the metal atoms at the surface, attached to ligand, have one electron localised in a covalent bond and thus contribute 2 valence electron each, which the other metal atoms

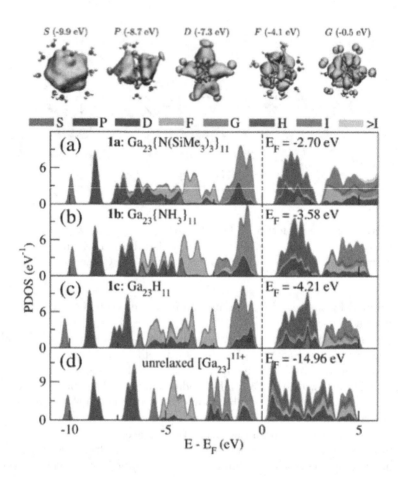

FIGURE 7.9: Top: The molecular orbitals of $Ga_{23}NH_2$, a model for $Ga_{23}\{N(SiMe_3)_2\}_{11}$, showing the nodal plane character of the labelled spherical harmonics. Bottom: The density of states of $Ga_{23}\{N(SiMe_3)_2\}_{11}$, and of its simplified variants, with the full ligand replaced by ammonia or hydrogen, or completely removed.

contribute three. Thus there are 11 divalent, and 12 trivalent metal atoms in the cluster, summing to the superatomic shell closing of 58 electrons. It is also worth noting that this analysis shows more variation in the properties of the atoms in the hydrogen-protected cluster; this demonstrates a reduced symmetry of the cluster, and thus that hydrogen as a ligand does not confer the same stability as the amine molecules used in experiment, as we would expect.

The above analysis can be further confirmed through the calculation of the Electron Localisation Function (ELF), which is a measure of the likelihood of

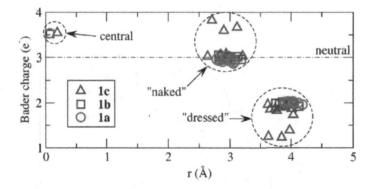

FIGURE 7.10: The Bader charges of the different gallium atoms in clusters protected by ligands of different complexity are plotted, based on their proximity to the surface of the cluster. Reproduced from Ref. [5] with permission from the PCCP Owner Societies.

finding an electron in the neighborhood space of a reference electron located at a given point and with the same spin:

$$D(\mathbf{r}) = \tau(\mathbf{r}) - \frac{(\nabla\rho(\mathbf{r}))^2}{4\rho(\mathbf{r})} \tag{7.3}$$

which can be compared to the known value for a uniform electron gas:

$$D^0(\mathbf{r}) = \frac{3}{5}(6\pi^2)^{2/3}(\rho(\mathbf{r}))^{5/3} \tag{7.4}$$

to give a parameter that varies from 1 to zero, with a value of 0.5 equating to the uniform electron gas:

$$ELF(\mathbf{r}) = \frac{1}{1 + \left(\frac{D(\mathbf{r})}{D^0(\mathbf{r})}\right)^2}. \tag{7.5}$$

Nitrogen based ligands are not the only kind that stabilise gallium clusters: indeed silicon based ligands – silanes – have been widely used. The $Ga_{22}[Si(SiMe_3)_3]_8$ cluster is a good example [6]. In Fig. 7.11, the structure of this cluster is given, calculated with the more complex silane ligand replaces by the SiH_3 molecule, which retains the same bonding mechanism of the Si atom to the surface Ga atoms. An isosurface of the ELF function has here been plotted for a value of 0.7, indicating relative localisation. The isosurface centred on the bond between the Ga and Si atoms confirms, indeed, the covalency of the bond and the corresponding localisation of one of the valence electrons of the Ga atom.

A final comment is worth making on the importance of electron shell models for ligand protected clusters. Once these clusters started to be analysed

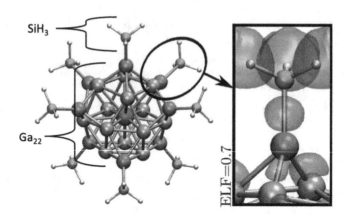

SiH$_3$

Ga$_{22}$

ELF=0.7

FIGURE 7.11: The structure of Ga$_{22}$[SiH$_3$]$_8$, and the ELF analysis of its ligand shell. Reproduced from Ref. [5] with permission from the PCCP Owner Societies.

in this way, it became clear that the superatomic concept was much more generally applicable than had been previously understood: it seems to be everywhere! However, consideration of the reasons for this reveal that an analogy of the anthropic principle is at work: the reason that so many experimentally synthesised system exhibit superatomic shell structure is that it is the clusters of these specific sizes that are most thermodynamically stable, and that will therefore naturally occur for a given combination of metal atoms and protective ligands.

Bibliography

[1] M. Walter, J. Akola, O. Lopez-Acevedo, P. D. Jadzinsky, G. Calero, C. J. Ackerson, R. L. Whetten, H. Grönbeck, and H. Häkkinen. A unified view of ligand-protected gold clusters as superatom complexes. *PNAS*, 105(27):9157–9162, 2008.

[2] D. Mollenhauer and N. Gaston. Phosphine passivated gold clusters: How charge transfer affects electronic structure and stability. *Physical Chemistry Chemical Physics*, 18(43):29686–29697, 2016.

[3] H. Schnöckel. Structures and properties of metalloid Al and Ga clusters open our eyes to the diversity and complexity of fundamental chemical and physical processes during formation and dissolution of metals. *Chemical Review*, 110(7):4125–4163, 2010.

[4] P. A. Clayborne, O. Lopez-Acevedo, R. L. Whetten, H. Grönbeck, and H. Häkkinen. The $Al_{50}Cp^*_{12}$ cluster - A 138-electron closed shell (L = 6) superatom. *European Journal of Inorganic Chemistry*, (17):2649–2652, 2011.

[5] D. Schebarchov and N. Gaston. Throwing jellium at gallium—a systematic superatom analysis of metalloid gallium clusters. *Physical Chemistry Chemical Physics*, 47(13): 21109–21115, 2011.

[6] A. Schnepf, E. Weckert, G. Linti, and H. Schnöckel. $Ga_{22}[Si(SiMe_3)_3]_8$: The largest atom-centered neutral main group metal cluster. *Angewandte Chemie; International Edition*, 38(22):3381–3383, 1999.

CHAPTER 8

Beyond Simple Superatoms

In Chapter 5, we discussed the simplest metal clusters possible: those prepared in the gas phase, from a range of different elements. We then complicated matters in Chapter 6, by introducing non-metal cluster based superatoms, and in Chapter 7, with the concept of ligand-protection, which, while it complicates the theoretical description and analysis considerably, in nonetheless necessary for the practical synthesis and manipulation of these types of species. However, in both these cases we have focused on systems where the stability of the cluster, necessary for it to have been synthesised in the first place, is due to an electronic shell closing that results, effectively, in superatomic mimics of noble gas atoms.

The identification of Al_{13} as a superatomic halogen – requiring only one additional electron for a closed shell, and thus strongly reactive with systems from where it can borrow an electron – is an indication that indeed, with an appropriate choice of metal and number of atoms, we might be able to design superatoms not to be inherently stable, but to have reactivity due to an open shell electronic structure. This is one particular context in which the history of molecular superalkali and superhalogen species provides a complementary view of what is possible to that formed by the study of metal cluster superatoms.

There are several different ways in which reactive superatomic character might be designed into metal clusters, and different pieces of evidence for what kind of bonding might be possible. In this chapter, we will start by considering the doping of superatoms, to tune their electronic structure as desired. We will then discuss the early transition metals, for which it has already been shown that the d-electrons can delocalise, leading to their participation in an electronic structure that may prefer to remain open shell due to magnetic interactions. This suggests the possibility of mimicking the chemistry and magnetic properties of the transition metal elements, using superatoms. Finally, we will consider what evidence there is in the literature for bonding between superatomic cluster units.

DOI: 10.1201/b23295-8

8.1   TUNABLE SUPERATOMS

One of the most exciting differences between atoms and their nanoscale analogues, is the ability we have, in principle, to modify the structure of a superatom through chemical substitution. The use of dopants to modify the electronic structure of bulk materials – in particular to enhance the properties of semiconductors – is a mature technology, and it is only natural that the idea soon arose to modify superatomic clusters in similar ways.

One of the most extensively studied superatoms, in particular in terms of its substitution chemistry, is the thiolate-protected Au_{25}. The metallic cluster core, ignoring the gold atoms that form part of the structure of the ligand shell, consists of an Au_{13} unit of icosahedral symmetry, which is approximately a sphere, with one atom at the centre. While substitution of both the ligands and metal atoms has been experimentally explored [1], the metal atom dopants are what are of most interest here, with relevance to the tunability of the electronic structure [2]. This structure is presented in Fig. 8.1.

For example, replacing a single gold atom by a platinum atom, which is from group ten in the periodic table rather than group 11, leads to a reduction in the overall electron count by 1 electron. Such substitutions have been reproduced with a number of different metal elements, in particular palladium, platinum, gold, and silver, but also cadmium and mercury, which in contrast to platinum and palladium, add an additional electron. Based on the experimental evidence, it is possible to calculate the electronic structure of these clusters using DFT, and compare to the original gold cluster with a single electron added or subtracted, in order to confirm the tunability of these superatoms [2].

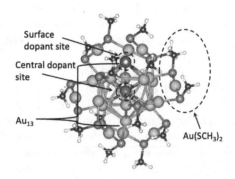

FIGURE 8.1: The structure of the tholate-protected Au_{25} cluster, showing the core cluster of 13 gold atoms surrounded by ligands. Two positions for dopants are possible within this core; the central site (shaded dark) and the surface of the 13 atom icosohedron (light grey).

In Fig. 8.2, the density of states of a doped Au_{25} cluster, that is thiolate protected, is presented. This demonstrates the essentially tunable nature of these species. Replacing one gold atom with a platinum atom means that there is one less valence electron delocalised over the metallic cluster core. Thus, the neutral cluster now has an open shell, with the $1P$ shell being the highest occupied, but now split into two peaks, with five electrons occupying the $1P$ state below the Fermi energy, and a single unoccupied state now visible above the Fermi energy. Adding an electron to this cluster recovers the closed shell $1P$ state known from the unpoded Au_{25} cluster.

However, not all clusters are equally tunable; the size of the dopant atom has an effect of the preferred location of the dopant, with some clusters having prepared (such as with the copper, cadmium, and mercury dopants) that have the dopant in a surface position within the Au_{13} icosahedron. If the size mismatch is even greater, then the doped cluster may never be synthesised, explaining that not all metal atoms can be doped into a given cluster in this way. On the other hand, so long as the dopants are compatible, it may be possible to introduce multiple dopant atoms into a metal cluster, or even combine dopants of even types, such as with the synthesis of gold clusters doped with both silver and palladium, or silver and platinum.

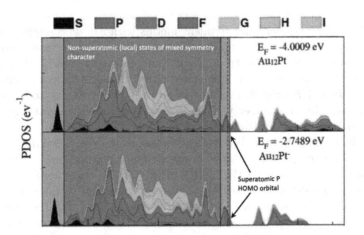

FIGURE 8.2: The PDOS of the tholate-protected Au_{25} cluster with a Pt dopant atom at the centre, demonstrating that in this case the neutral cluster is an open shell superatomic mimic of a halogen atom, and that the negatively charged cluster has the usual closed shell electronic structure. Reproduced from Ref. [2] with permission.

8.1.1 Magnetic Dopants

One of the many promises suggested by dopants, however, is the ability not only to add or subtract a given number of delocalised valence electrons, thereby modifying the electronic shell structure, but by doping in a transition metal with a number of unpaired d-electrons, creating a superatom that has a magnetic moment. For example, a superatom composed of alkali metal atoms could have a transition metal introduced, creating a superatom in which the magnetic moment is sensitive both to the size of the cluster, and to the type of metal used. Various studies have examined different types of such species, such as magnesium clusters doped with iron [3], or calcium clusters doped with a range of transition metal species [4]. In the majority of these cases, the cluster itself remains chemically unreactive as a superatomic shell-closing is still energetically preferred, while the magnetic moment of the cluster is due to strongly localised d-electrons that do not participate in the superatomic electronic shells. However, recent work analysing the interaction between superatomic and local electron density in transition metal clusters suggests that it is also possible to obtain transition metal based clusters that are inherently open shell, as was discussed in Chapter 5 [5], or in which the magnetic properties are tunable as a function of size [6].

8.2 OPEN SHELL MAGNETIC SUPERATOMS

In Chapter 5, in introducing the case of transition metals in the superatomic model, it was mentioned that trivalent metals with a s^2d^1 configuration can let the d-electron delocalise sufficiently to participate in electron shells, but that the ordering of the electronic shells (and subshells) can be strongly affected by the symmetry of atomic cores within the cluster, as certain orientations of D and F orbitals can be energetically preferred due to a breaking of degeneracy due to symmetry. This scenario is depicted schmatically in Fig. 8.3. There are two major consequences of this.

Firstly, it suggests that these transition metals may be useful in the design of superatomic clusters that have an open shell configuration, and that this might lead to them having the ability to interact in quite novel ways. Secondly, it suggests that these metals will be very sensitive to the environment they are placed in, as the orientation of the atomic d-orbitals relative to the cluster framework can have a significant impact on the degeneracies of the superatomic orbitals, and thus to the appropriate electron counting rules.

This second point is perhaps demonstrated most readily by taking open shell transition metals and placing them in a framework that is designed to enhance their ability to delocalise their valence electrons. A useful point of comparison is obtained by taking metals from group 6 of the periodic table – each with 6 valence electrons – and comparing their behaviour. The lighter two elements, chromium and molybdenum, have a s^1d^5 configuration as this allows for all the valence atomic orbitals to be singly occupied, meaning that

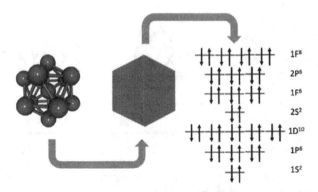

FIGURE 8.3: The superatomic states of an icosahedral system with 40 valence electrons. The angular momentum labels of the states are shown on the right hand side of the image. While the lower energy shells are fully occupied, the higher energy states formed from atomic d-electrons remain open shell. Reproduced from Ref. [5] with permission from the PCCP owner societies.

the atom has 6 aligned unpaired electrons, maximising its magnetic moment. On the other hand, in the third row of the transition metals, the d-orbitals have a significantly greater extent, shielded as they are from the nucleus by the many core electrons, including for the first time, f-electrons. This means that the atom adopts a s^2d^4 configuration, with the s-electrons paired and therefore only four unpaired d-electrons.

When placed in an idealised superatomic environment, these atoms therefore behave differently. In Fig. 8.4, the PDOS are plotted for clusters of 14 sodium atoms that contain a Cr, Mo, or W atom at the centre [7]. Each sodium atom contributes a single valence electron, meaning that if the transition metal atoms can be persuaded to delocalise all their 6 valence electrons, the cluster will have a superatomic closed shell. For each cluster, the calculations have been repeated with a range of density functionals; PW91, PBE, mPW, BLYP, and BP86 are all examples of GGA functionals, while PBE0, mPW1, and B3LYP are all examples of hybrid density functionals that include exact exchange, increasing the relative localisation of the d-electrons.

It is straightforward to note that the electronic structure of the Cr and Mo containing clusters is open shell in all cases, driven by the tendency of the d-electrons to remain unpaired and aligned. Thus while the whole cluster has a superatomic character predetermined by the Na atoms, the superatom is robustly open shell and magnetic. The case of tungsten, however, as might have been expected, is different. In the case of two of the hybrid DFT calculations, the same result is found as for the lighter transition metals: an open-shell superatom. However, without exact exchange, the d-orbitals are sufficiently

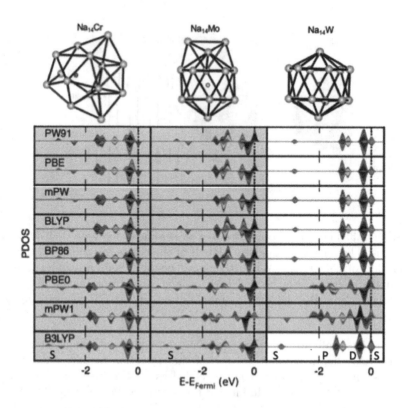

FIGURE 8.4: The PDOS of Na$_{14}$ superatoms with a Cr, Mo, or W atom at the centre. Above are shown the optimised structures of these clusters. Reproduced from Ref. [7] with permission from the PCCP owner societies.

delocalised to participate fully in the superatomic shells, leading to all 20 electrons found in a closed-shell $1S^2 1P^6 1D^{10} 2S^2$ electronic structure.

This result is replicated for one of the calculations where exact exchange was also included. While in some ways this demonstrates the usual frustrations that can we found in using DFT, where determining the 'right' choice of functional is, *a priori*, an impossible task, it also illustrates that the choice of electronic structure in this cluster is very sensitive to small changes. This leads to the prospect of such a cluster being able to be switched from magnetic to closed shell by small changes in its surrounding environment. This is the truly exciting result of studying these types of clusters, as the ability of transition metals to adapt their electronic configuration in response to their environment is what makes them so useful in both technological (magnetic materials) and biological (enzymes) applications. Here we see that this may still be possible for an entire superatom.

8.3 SUPERATOMIC MOLECULES

The prospect of superatoms interacting with each other in ways that could be exploited for the design of nanostructured materials is already attested to, experimentally, by the observation of numerous species that can be best understood electronically, as superatomic molecules. These start with the familiar and well-studied ligand-protected gold clusters that were discussed in Chapter 6. In Fig. 8.5, the electronic states of a Au_{23}^{9+} core are compared to those of the fluorine molecule, demonstrating that superatomic orbitals based within each moiety of the Au_{23}^{9+} can interact in the same ways as atomic orbitals between individual atoms. The Au_{23}^{9+} core exists within the $Au_{38}(SR)_{24}$ molecule that has been synthesised by the same means as the original ligand-protected gold superatoms, but with only 14 delocalised electrons left after 24 valence electrons are localised by the ligand shell, it was a mystery why it should be a preferentially stable species until the electronic structure was understood to be that of a closed shell molecule [8].

Subsequently, specific efforts have been made to synthesise superatomic molecules, starting from superatomic building blocks. In this case, rather than obtaining a fused metallic core from a single synthesis, the clusters are formed and the ligands are designed to facilitate interconnections between clusters.

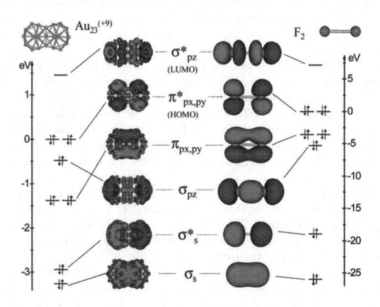

FIGURE 8.5: The electronic states and orbitals of the Au_{23}^{9+} core of a ligand protected $Au_{38}(SR)_{24}$ cluster are compared to those of fluorine, demonstrating it to be a superatomic molecule. Figure reproduced with permission [8].

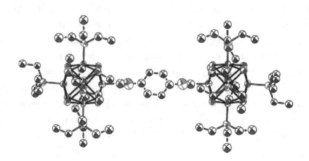

FIGURE 8.6: The structure of a ligand connected superatomic molecule consisting of two Co_6Se_8 cluster cores, connected by CNC_6H_4NC ligands and protected externally by phosphine ligands. Figure reproduced with permission [9].

Such an approach may be particularly promising for connecting two clusters while preserving their electronic structure, such as may be required to enable two magnetic species to interact; an example of such a recently synthesised structure [9] is given in Fig. 8.6.

As a final example, superatomic molecules have been designed to contain the essential characteristics of a material useful for specific application. Photovoltaic materials rely on a p-n junction, effectively an interface where a light induced electron-hole pair, after excitation, may be preserved for a finite time through separation by an electric field. Fusing two superatomic clusters that have ligands to accept or donate electron density on opposite sides has been shown to lead to a strong internal electric field that amounts to such a junction [10].

In Fig. 8.7A-B, the fused metal-chalcogenide $Re_6S_8Cl_2(L)_4$ clusters are shown, with donor PMe_3 ligands and acceptor CO ligands on the opposite sides of the fused clusters. Fig. 8.7A shows an electron density isosurface for the molecule when an electron is added to it, showing that the electron is accepted by one half of the molecule, while Fig. 8.7B shows the isosurface when an electron is removed, showing that the hole is located on the other half of the superatomic molecule. In Fig. 8.7D the DOS is shown for the different parts of the molecule (as outlined in Fig. 8.7C); this demonstrates that the electronic levels undergo shifts analogous to band bending in traditional p-n junctions.

In summary, the concept of superatoms is extendable well beyond its original use in explaining the stability of individual superatomic clusters. Both the demonstration of open shell species accessible through doping or the use

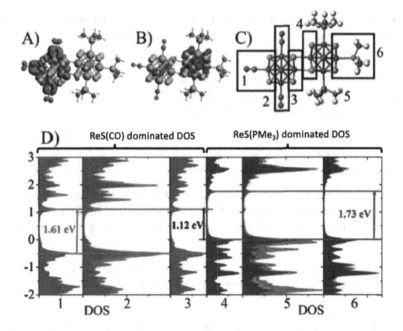

FIGURE 8.7: The electronic structure of a superatomic molecule that can function as a p-n junction. A) the structure of the anionic cluster, with the isosurface showing the excess electron density on the acceptor ligands. B) the cationic cluster. C) the regions of the molecule for which the localised DOS is plotted in D. Figure reproduced with permission [10].

of magnetic transition metals, and the realisation of superatomic molecules, illustrates a vast vista of perspectives that are opening up for the design of nanostructured materials. We will discuss some early realisations of these in the following chapter.

Bibliography

[1] Y. Negishi, W. Kurashige, Y. Niihori, and K. Nobusada Toward the creation of stable, functionalized metal clusters. *Physical Chemistry Chemical Physics*, 15:18736–18751, 2013.

[2] J. Schacht and N. Gaston. From the superatom model to a diverse array of super-elements: A systematic study of dopant influence on the electronic structure of thiolate-protected gold clusters. *ChemPhysChem*, 17(20):3237–3244, 2016.

[3] V. M. Medel, J. Ulises, S. N. Khanna, V. Chauhan, P. Sen, and A. Welford Castleman. Hund's rule in superatoms with transition metal impurities. *PNAS*, 108(25):10062, 2011.

[4] V. Chauhan, V. M. Medel, J. U. Reveles, S. N. Khanna, and P. Sen. Shell magnetism in transition metal doped calcium superatom. *Chemical Physics Letters*, 528:39–43, 2012.

[5] J. T. A. Gilmour and N. Gaston. On the involvement of d-electrons in superatomic shells: The group 3 and 4 transition metals. *Physical Chemistry Chemical Physics*, 21:8035–8045, 2019.

[6] J. T. A. Gilmour, L. Hammerschmidt, J. Schacht, and N. Gaston. Superatomic states in nickel clusters: Revising the prospects for transition metal based superatoms. *Journal of Chemical Physics*, 147(15), 2017.

[7] J. T. A. Gilmour and N. Gaston. Design of superatomic systems: exploiting favourable conditions for the delocalisation of d-electron density in transition metal doped clusters. *Physical Chemistry Chemical Physics*, 22:18585–18594, 2019.

[8] L. Cheng, C. Ren, X. Zhang, and J. Yang. New insight into the electronic shell of $Au_{38}(SR)_{24}$: a superatomic molecule. *Nanoscale*, 5:1475–1478, 2013.

[9] A. M. Champsaur, A. Velian, D. W. Paley, B. Choi, X. Roy, M. L. Steigerwald, and C. Nuckolls. Building diatomic and triatomic superatom molecules. *Nanoletters*, 16(8):5273–5277, 2016.

[10] A. C. Reber, V. Chauhan, D. Bista, and S. N. Khanna. Superatomic molecules with internal electric fields for light harvesting. *Nanoscale*, 12:4736–4742, 2020.

CHAPTER 9

Superatomic Assemblies

We started our discussion of superatoms by motivating their study as being due to the finite number of elements available to us for the construction of materials. We have since explored the additional advantages that superatoms can provide – their tunable structures, their ability to be joined together in specific geometries through ligands, and the extreme properties that can be designed into them, such as in the case of superhalogens that have higher electron affinities than actual halogens. In this chapter, we will turn our attention to the kinds of materials that have been made experimentally, and studied with the use of the superatomic concept. We'll then develop the analysis techniques that were introduced in Chapter 4 for individual metal clusters – in particular, the concept of a density of states, that can be used to identify electronic shell structure – into forms that are necessary for the description of extended, periodic systems. To do this we'll recap how a band structure is formed in a periodic system, and how electronic properties can be understood on the basis of such band structures.

9.1 EXPERIMENTAL EVIDENCE OF SUPERATOMIC ASSEMBLIES

If a superatom is to be used to create a nanostructured material, it needs to be assembled into a crystal structure. One of the clearest demonstrations of this happening to date, is the co-crystalisation of Co, Cr, and Ni based clusters together with C_{60} molecules, known also as buckyballs. The structures of the various clusters are shown in Fig. 9.1a, and the structures of the resulting binary assemblies are in part b.

DFT calculations have been performed to test the extent to which these assemblies may be understood as superatomic. That is: do they exhibit properties that could be understood to arise from some form of bonding interaction between the metal based clusters, and the buckyballs?

The first part of this analysis is presented in Fig. 9.2, for the case of the Co-based cluster of the three mentioned above. The cohesive energy of a solid

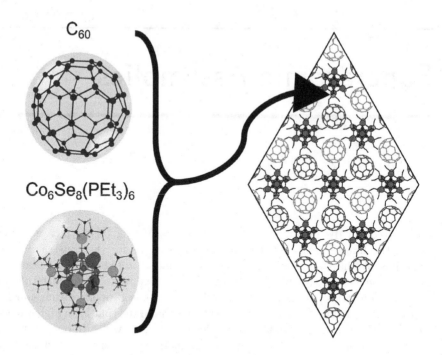

C$_{60}$

Co$_6$Se$_8$(PEt$_3$)$_6$

FIGURE 9.1: The lattice structure of a superatomic assembly is shown. At left, the two building blocks of the assembly are shown; at right the superatomic crystal structure is shown.

is calculated as the difference between the energy of the composite material and its contributing parts. Thus for an assembly of a C$_{60}$ buckyball and a superatomic cluster, which each have energies E(C$_{60}$) and E(SA), we can write the cohesive energy as:

$$E(\text{cohesive}) = E(\text{unit cell}) - [nE(C_{60}) + mE(SA)] \qquad (9.1)$$

where in each unit cell required to construct the periodic lattice, there are n buckyballs and m superatomic clusters, and thus we are left with only the energy of interaction. For a solid to be stably bound, this energy should be by definition negative.

The cohesive energies in Fig. 9.2 can be seen to vary considerably depending on the density functional chosen. This can be a frustration of DFT, that such variations occur, but in this case the differences can give us some insight into the nature of binding, due to the well known characteristics of these particular functionals. Superatomic assemblies seem to be better described by the LDA than by a GGA (here the specific functional is denoted PBE), despite the fact that the GGA should in principle be more accurate. This is a type of error cancellation often seen for systems for which the correlation energy

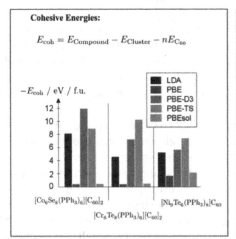
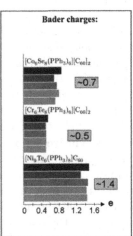

FIGURE 9.2: Comparison of the binding energy of the calculated structures of potential superatomic assemblies, as obtained through DFT calculations. Reproduced from Ref. [2] with permission from the PCCP owner societies.

contribution to binding is important. The LDA usually overbinds relative to experiment, but this overbinding is often approximately equal to the underestimation of the correlation energy of electrons that are not chemically bound, but that are interacting through long-range dispersion forces, which are also referred to as van der Waals interactions. These arise from the mutual interaction of instantaneous-dipoles created from the motion of charge, and are therefore not captured within any independent particle approximation: they are inherently non-local.

The -D3 and -TS corrections shown in this figure are both ways of correcting for the missing dispersion interaction, via a parametrised long range van der Waals force which goes as r^{-6}, which is the appropriate attraction between two electrostatic dipoles at long range. Comparison of the actual distances between superatoms and C_{60} molecules in the lattice show that these dispersion corrections reduce the volume of these types of solids from being overestimated by up to 20%, to being within a few percent of experimental values, verifying that the dispersion forces are responsible for most of the binding between clusters in these materials [2].

In combination with the calculated Bader charges, presented to the right hand side of Fig. 9.2 a), we can describe here one of the central paradoxes of the interactions between superatoms in the solid state. Because of their large size and stable molecular character, they do not interact strongly via covalent bonds. Instead, crystals can form that are held together primarily by 'weak' non-bonding interactions, such as the long-range dispersion force. However, once they are in sufficiently close proximity, the electronic structure of the superatoms becomes important in its own right, and – for structures in which the clusters are held in sufficient proximity by the dispersion force – they can interact electronically, via the transfer of charge from one cluster to another. It is this electronic interaction, even though it is decoupled from the binding force of the solid, that leads to the prospect of superatomic solids that have tunable electronic structure.

The DOS of the clusters in this assembly has also been calculated, but does not provide a conclusive description of the electronic states of the assembled material. The superatomic electronic shell structure is shown to exist, but not to dominate behaviour of the highest occupied orbitals that describe the interacting electrons, and thus not of the properties of the material. For a more in depth look at the interacting states in these superatomic crystals we will need to explore the band structures of these systems: the ways in which the geometric arrangement of the lattice influences the ability of electrons to transfer between subunits of the material.

9.2 THE BAND STRUCTURE OF SOLIDS

Before exploring the band structures of these superatomic solids, we will recap some key concepts and terms which will help in the interpretation of the data. Both the description the lattice structures themselves, and the Brillouin zone required for the interpretation of electronic energies in the solid state, are consistent with their usual descriptions – the only difference for superatomic solids is that the lattice spacings are effectively larger, as each unit cell must contain all the atoms associated with each of the superatoms required in the unit cell.

9.2.1 Crystal Structures

A crystal lattice is defined by a set of vectors which describe the borders of the *unit cell* of the crystal. The unit cell can therefore be tiled precisely in three dimensions to obtain a perfect infinite lattice which fills all space.

$$\mathbf{r}' = \mathbf{r} + u_1 \mathbf{a_1} + u_2 \mathbf{a_2} + u_3 \mathbf{a_3} \tag{9.2}$$

where u_1, u_2, u_3 are arbitrary integers. The angles between vectors $\mathbf{a_1}, \mathbf{a_2}, \mathbf{a_3}$ are labelled α, β, γ.

Not all shapes can be tiled perfectly in three dimensions - therefore there is a finite set of lattices which are possible. For our purposes in describing superatomic assemblies, there are exactly the same as described in any solid state textbook. Examples of objects that cannot be tiled perfectly include pentagons (in 2 dimensions) and icosahedra (in 3D). Despite the limited number of lattice types that are available, we can further complicate the structures of materials by allowing for multiple atoms to sit within a given unit cell. The position of an individual atom (labelled i) can be written:

$$\mathbf{r_i} = x_i\mathbf{a_1} + y_i\mathbf{a_2} + z_i\mathbf{a_3}. \tag{9.3}$$

It should be seen from the ability to put numerous atoms within a unit cell that there is no limitation on the size of the unit cell: we can tile a finite number of unit cells together to create a *supercell*. For example, by tiling $3 \times 3 \times 3$ cells together we can create a supercell with the same symmetry as the original but with 27 times as many atoms inside. This can be useful, for example, when we need to study the presence of defects or dopants in materials.

9.2.2 The Brillouin Zone

Experimentally, and as introduced in many textbooks, the use of the Brillouin Zone (and the need to define the reciprocal lattice) is motivated by the fact that we use diffraction to study the properties of crystals. However, there is another way to think about the need to work in momentum space, which is that the energies of the electrons in solids depend on their momentum within the crystal, and therefore on the symmetry of the space that they are moving through. Therefore, to get an accurate energy in a DFT calculation of a solid, we need to calculate the energy eigenvalues that each electron may have, as a function of the momentum that each electron may have.

We can define the axis vectors of the *reciprocal lattice* as

$$\mathbf{b_1} = 2\pi\frac{\mathbf{a_2} \times \mathbf{a_3}}{\mathbf{a_1} \cdot \mathbf{a_2} \times \mathbf{a_3}}; \quad \mathbf{b_2} = 2\pi\frac{\mathbf{a_3} \times \mathbf{a_1}}{\mathbf{a_2} \cdot \mathbf{a_3} \times \mathbf{a_1}}; \quad \mathbf{b_3} = 2\pi\frac{\mathbf{a_1} \times \mathbf{a_2}}{\mathbf{a_3} \cdot \mathbf{a_1} \times \mathbf{a_2}}. \tag{9.4}$$

If $\mathbf{a_1}, \mathbf{a_2}, \mathbf{a_3}$ are primitive vectors of the crystal lattice (e.g. describe the primitive cell) then the vectors $\mathbf{b_1}, \mathbf{b_2}, \mathbf{b_3}$ are primitive vectors of the reciprocal lattice. These vectors may be used to construct the *Brillouin zone*, which is the primitive cell in reciprocal space.

The first Brillouin zone contains within it all the possible values of momentum that affect electronic energies within the crystal. Certain high-symmetry points within the Brillouin zone are given specific labels, which will appear without further explanation on some of the figures; readers are advised to consult a specialist textbook for the relevant Brillouin zone diagrams and high-symmetry point definitions.

From our discussion of the free electron Schrödinger equation in Chapter 1, we learned that

$$m\mathbf{v} = \hbar\mathbf{k}. \tag{9.5}$$

For a given crystal lattice, we need to systematically sample the k-points (values of $\mathbf{k}$) within the Brillouin zone (which contains the full range of momentum values needed to describe energy dispersion): we do this by solving the Kohn-Sham equations in DFT, or the Schrödinger equation on a evenly

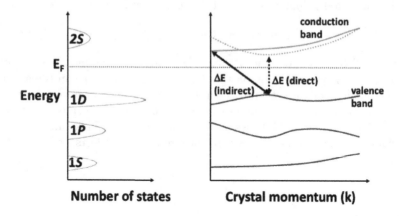

FIGURE 9.3: Schematic of the relationship between the density of states, introduced earlier (left) and the band structure in which the same energy eigenvalues are plotted as a function of momentum (right). The definition of a band gap is shown, depending on whether it involves a change of photon momentum (indirect, arrow to the minimum of the solid line conduction band shown) or no change of momentum (dashed conduction band).

spaced set of grid points within the Brillouin zone. In order to interpret the energy dispersion within the Brillouin zone, we plot the energy eigenvalues at a systematically generated set of k-points which trace out the connections between the high symmetry points of the Brillouin zone mentioned earlier. In effect, we are reducing 3D data to 1D for practicality.

Fig. 9.3 provides a schematic explantion of how such a band structure relates to the density of states described earlier. In essence, a density of states sums over all energies regardless of the atoms those states belong to. The spread of energies that results can be understood, in the solid state, as resulting from a change of the energy of a particular electronic state depending on the momentum it has within the crystal lattice. This depends in turn on the symmetry of the lattice itself, so on the positions of the atoms that determine if a lattice is hexagonal, or cubic, or something else. What this means, in the context of superatomic assemblies, is that by changing the lattice structure, the band structure can be modified in ways that affect the band gap – the critical energy value that determines the way a material can conduct electrons, or interact with light, and that thus determines so many of its eventual properties of interest.

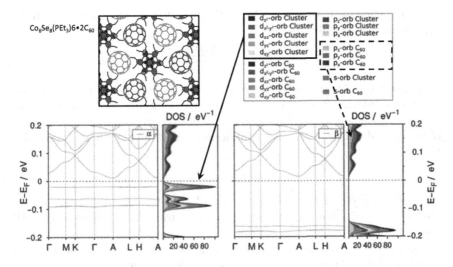

FIGURE 9.4: The band structure of the Co-based superatomic assembly is presented, together with the corresponding (projected) DOS. Adapted from Ref. [2] with permission from the PCCP owner societies.

9.3 THE BAND STRUCTURE OF SUPERATOMIC ASSEMBLIES

The band structures for the Co-based cluster assembly described in Fig. 9.1, is provided in Fig. 9.4. For each material, the spin up and spin down bands are plotted separately, at left and right.

The bands below the Fermi energy, that are occupied, are those of the metal cluster, and largely composed of d-electron density, which is why the energies of the spin up and spin down plots are different, as the d-electrons are naturally magnetic. This also explains why these bands are very flat – much less disperse – than in regular solid state materials (it is notable that the unoccupied bands above the Fermi energy that belong primarily to the C_{60} molecule are much more disperse. How can we understand this?

The simplest model of the band structure is based on understanding that electrons in solids behave in a way that can be seen as a perturbation of free electron behaviour, and thus of the free electron dispersion relation. If the band structure is known, we can expand the energy wave vector relation $E(k)$ in a one-dimensional Taylor series as

$$E(k) = E(0) + \frac{\partial E(k)}{\partial k}\bigg|_{k=0} k + \frac{1}{2}\frac{\partial^2 E(k)}{\partial k^2}\bigg|_{k=0} k^2 + \dots \qquad (9.6)$$

If we set $k = 0$ to be the position of the conduction band minimum, the gradient of $E(k)$ will be zero at $k = 0$. From this it follows that the lowest order expression is

$$E(k) = E(0) + \frac{\hbar^2 k^2}{2m^*} \qquad (9.7)$$

where the effective mass is a scaling of the free electron mass according to the 'weight' of the crystal potential:

$$\frac{1}{m^*} \equiv \frac{1}{\hbar^2} \frac{\partial^2 E(k)}{\partial k^2}. \tag{9.8}$$

In the more general three-dimensional case the conduction band may have mulitple minima, often at $\mathbf{k} = 0$ (called the Γ point), and others at high symmetry points in the Brillouin zone, which vary according to the symmetry of the crystal lattice. If the first conduction band minimum $E(\mathbf{k})$ is described by

$$E(\mathbf{k}) = \frac{\hbar^2 k^2}{2m^*} \tag{9.9}$$

then we can say that the effective mass m^* is isotropic. This corresponds to the existence of a spherical constant energy surface in $\mathbf{k}$ -space around the band minimum.

In a cubic crystal, we can expand the above expression into three dimensions and define longitudinal and transverse electron masses, m_l^* and m_t^*:

$$E(\mathbf{k}) = \frac{\hbar}{2} \left(\frac{k_\ell^2}{m_\ell^*} + \frac{k_t^2}{m_t^*} + \frac{k_t^2}{m_t^*} \right). \tag{9.10}$$

This general approach is called the parabolic energy band approximation, in recognition of the assumption that the band minimum is quadratic. It can also be applied to the maxima of the valence band, in which case the effective masses obtained are those that determine the behaviour of holes – electron vacancies – in the valence band.

How does the effective mass matter? One key consideration is that the charge carrier (electron or hole) velocity is inversely proportional to the effective mass:

$$\mathbf{v} = \frac{e\tau}{\|m^*\|} \cdot \mathbf{E}, \tag{9.11}$$

with τ defined as the mean free time between ionic collisions in the Drude model of conductivity. In general therefore, the electronic functionality of a material will be greater, when the bands are more disperse.

9.4   ELECTRONIC PROPERTIES OF CLUSTER ASSEMBLIES

Notably, the assemblies for which we have discussed the structures above demonstrate both electronic transport and magnetic properties that arise from interactions of the constituent clusters. Experimental evidence of the electronic transport and magnetic properties of the three Co, Cr, and Ni-based cluster assemblies are outlined, to demonstrate not only that these can result from interactions between assembled clusters, but that the nature of these properties depends sensitively on the properties of the individual clusters. This suggests that we there are meaningful interactions represented in the band structure, to which we can fit simple models – for example Boltzmann theory, discussed below – in order to compare with experiment.

9.4.1   Boltzmann Theory: Application to Superatomic Solids

The electronic transport properties of these superatomic solids are straightforwardly extractable from the band structures – the relatively flat bands mean that the size of the band gap is the main quantity of interest. The experiment found good electrical conductance for both compounds of the hexagonal symmetry, with a semiconducting behaviour with Arrhenius activation energies E_a of about 150 meV and 100 meV, respectively. The calculated band structure is in agreement, both compounds are small-gap semiconductors with band gaps of about 50–100 meV depending on the spin of the bands, and the DFT functional. Because DFT functionals generally underestimate the band gap (due to the self-energy problem), we have good overall agreement with the experimental activation energies.

However, because the energies are so dependent on the spin channel, we can expect to observe interesting magnetic properties. Indeed, the experiment found a different magnetic response for each of the Cr-based and Co-based cluster assemblies. In an attempt to capture the paramagnetic response we can consider a rigid-band model without spin-polarisation for the two compounds 1 and 2.

The rigid-band approximation (RBA) assumes that changing the temperature, or doping a system, does not change the band structure. Within this approximation the carrier concentration in a semiconductor can be obtained from the density of states $n(\epsilon)$:

$$n(\epsilon) = \int \Sigma_b \delta(\epsilon - \epsilon_{b,\mathbf{k}}) \frac{d\mathbf{k}}{8\pi^3} \tag{9.12}$$

The index b is now used to make it clear that the sum runs over the bands. The deviation from charge neutrality provides the carrier concentration:

$$c(\mu, T) = Q - \int n(\epsilon) f^0(\epsilon; \mu, T) d\epsilon \tag{9.13}$$

Here Q is the nuclear charge and f^0 the Fermi distribution function which captures the temperature dependence. μ is the chemical potential, which is to say it is a measure of the doping of the system: if it is equal to an energy above the Fermi energy, in the conduction band, that means you have an n-type material (with electrons free to move in the conduction band), while if μ is less that the Fermi energy, and lies within the valence band, that means that there are holes within the valence band, and thus you have a p-type material (where the charge carriers are positively charged).

The simplest way to implement these equations is with the use of a parabolic band model:

$$\epsilon(\mathbf{k}) = \frac{\hbar^2 \mathbf{k}^2}{2m^*} \tag{9.14}$$

using the effective mass m^*. This allows us to obtain simple analytic expressions for the density of states and the transport coefficients:

$$n(\epsilon) = \frac{1}{4\pi^2} \left(\frac{2m^*}{\hbar^2}\right)^{3/2} \epsilon^{1/2} \tag{9.15}$$

and

$$\sigma(\epsilon) = \frac{1}{3\pi^2} \frac{\sqrt{2m^*}}{\hbar^3} \tau \epsilon^{3/2}. \tag{9.16}$$

The most illustrative comparison between theory and experiment has been obtained for the magnetic susceptibility of these superatomic assemblies. We can calculate the inverse of the Pauli magnetic susceptibility as:

$$\chi(T;\mu) = \mu_0 \mu_B^2 \int n(\epsilon) \left[\frac{\partial f_\mu(T;\epsilon)}{\partial \epsilon}\right] d\epsilon \tag{9.17}$$

with respect to the chemical potential μ at a selected temperature. For comparison to experimental results this has been done for three values of μ that correspond to sensible n- and p-dopings, respectively. μ_0 and μ_B are the vacuum permeability and Bohr magneton constants, respectively.

The results of this calculation, within a semiclassical approximation and a constant relaxation time approximation, as above, are given in Fig. 9.5.

9.5   SUPERATOMIC CHARACTER IN THE SOLID STATE

A different application of the superatomic model is not to the design of new materials, but to the interpretation of the structure of existing materials. This is the case of work looking to understand the behaviour of superionic conductors based on cluster ions [3].

Here, by exploring a set of lithium-rich antiperovskites (LRAPs) composed of cluster ions (Li_3O^+/Li_3S^+ and $BH_4^-/AlH_4^-/BF_4^-$), the authors report crystalline materials that have estimated high conductivity of 10^{-2} S/cm and activation energies around 0.2 eV. The materials are composed of cluster cations and cluster anions, which are known as superalkalis and superhalogens, respectively. The superalkalis have ionisation potentials (IPs) smaller than those of alkali elements, and the superhalogens have vertical detachment energy (VDE) larger than that of halogen elements. By partially replacing the superhalogen ion BF_4^- with chlorine, the mixed phase material, $Li_3S(BF_4)_{0.5}Cl_{0.5}$, improves conductivity to over 10^{-1} S/cm at RT and has an activation energy as low as 0.176 eV.

The reason for this is again the decoupling of structural stability from actual chemical bonds, by manipulating the size of the building blocks. By putting the elementary halogen and the large superhalogen together in the same structure, the lattice is required to accommodate the superhalogen (BF_4^-), but this leaves a lot of extra space around the positions of the Cl^- atom. The Li^+ ion has, consequentially, a lot of extra space with unusually

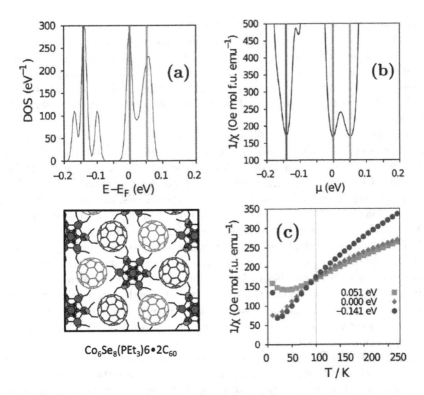

FIGURE 9.5: Pauli magnetic susceptibility as obtained from semi-classical Boltzmann transport theory for $[Co_6Se_8(PEt_3)_6][C_{60}]_2$. (a) The DOS; (b) the magnetic susceptibility with respect to the chemical potential; (c) temperature dependence of the magnetic susceptibility. The structure of the assembly is also shown. Adapted from Ref. [2] with permission from the PCCP owner societies.

large space to migrate in the halogen-containing cells, while keeping the original conductivity for the cells containing the superhalogens.

On the basis of this identification of existing families of materials as being superatomic in nature, design principles can be used to modify the electronic structure of technologically important materials based on the superatomic concept. A recent example is the suggestion that superalkali metal dimers can be used to dope perovskites that are known for their use as solar cell materials, and that for which the band gap is a key parameter of particular interest [4].

In this work, superalkali cations were introduced to replace the Cs^+ cation in a common perovskite, $CsPbBr_3$. Cation substitution has previously been used to modify perovskites in a number of ways, with there being two key factors that affect the band gap: firstly, the size of the cation can stabilise or destabilise the crystal structure of the perovskite away from its simplest cubic form. The lattice structure influences the band gap, as outlined earlier in this chapter; in addition, however, the electron donating character of the cation has a significant influence on the postion of the electronic states. Bimetallic superalkali species, LiMg, NaMg, LiCa, and NaCa, have a strong tendency to lose one electron to achieve a closed-shell cation, and were considered to be the right size to be substituted for Cs without destabilising the structure excessively.

All these compounds are shown to be able to form stable perovskites, with, however, a structural distortion to a tetragonal, rhombohedral, or orthorhombic symmetry. This superalkali cation causes a tilting of the $PbBr_6$ octahedra to optimise bonding between the halogen atoms and the superalkali dimer. The significantly less electropositive superalkalis, compared to cesium, result in lower band gaps. The smaller of the superalkali species, specifically $LiMg^+$ and $NaMg^+$, form semiconductors with a bandgap from 0.25 to 1.54 eV. The larger superalkalis result in metallic systems with overlapping conduction and valence bands, in particular due to the extra electronic states introduced around to the Fermi level, which arise from the formation of alkali earth metal states at the top of the valence band. A key takeaway from this work is that with the right substitution of superalkali species, it is possible to move the bandgap to within the optimal range (i.e., 1.1-1.4 eV) for single-junction solar cells.

Figure 9.6 demonstrates the sensitivity of the band structure to the orientation of the superalkali dimer within the perovskite lattice, for the case of the simplest (tetragonal) lattice distortion. This sensitivity also provides additional insight into the variety of mechanisms by which the electronic properties of superatomic materials may be tuned, using these simple concepts.

9.6 SUMMARY

Since the first proposal of the superatomic concept as something more than a simple application of the jellium model to bare metal clusters, the prospect of applying these ideas to the design of real functional materials has been on the horizon. In this text I have attempted to introduce some of the most significant ways in which the tunability of superatomic clusters has been demonstrated and applied, and I hope I have been able to point to some of the most interesting emerging areas in materials design that employ the principles that have emerged from the study of superatoms and their interactions.

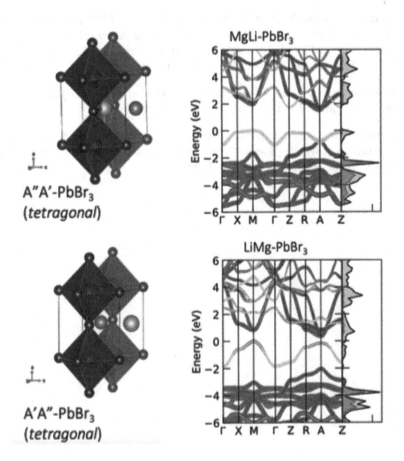

FIGURE 9.6: Left: the two different sites for the Mg and Li atoms in the MgLi dimer to be substituted are shown. Right: the band structures calculated at the HSE06 level of theory are shown, demonstrated the strong sensitivity to superalkali orientation within the perovskite lattice. Figure reproduced with permission [4].

Bibliography

[1] X. Roy, C.-H. Lee, A. C. Crowther, C. L. Schenck, T. Besara, R. A. Lalancette, T. Siegrist, P. W. Stephens, L. E. Brus, P. Kim, M. L. Steigerwald, and C. Nuckolls. Nanoscale atoms in solid-state chemistry. *Science*, 341(6142): 157–160, 2013.

[2] L. Hammerschmidt, J. Schacht, and N. Gaston. First-principles calculations of the electronic structure and bonding in metal cluster–fullerene materials considered within the superatomic framework. *Physical Chemistry Chemical Physics*, 18: 32541–32550, 2016.

[3] H. Fang and P. Jena. Li-rich antiperovskite superionic conductors based on cluster ions. *PNAS*, 114: 11046–11051, 2017.

[4] C. Sikorska and N. Gaston. Bimetallic superalkali substitution in the $CsPbBr_3$ perovskite: Pseudocubic phases and tunable bandgap. *Journal of Chemical Physics*, 155: 174307, 2021.

Index

Printed in the United States
by Baker & Taylor Publisher Services